Disease Management in Cocoa

Disease Management in Cocoa

Comparative epidemiology of witches' broom

Edited by S.A. Rudgard, A.C. Maddison and T. Andebrhan
International Office of Cocoa, Chocolate and Sugar Confectionary

SPRINGER-SCIENCE+BUSINESS MEDIA, B.V

First edition 1993

Originally published by Chapman & Hall in 1993
Softcover reprint of the hardcover 1st edition 1993
ISBN 978-94-010-4943-6 ISBN 978-94-011-2126-2 (eBook)
DOI 10.1007/978-94-011-2126-2

A catalogue record for this book is available from the British Library

Library of Congress Cataloging-in-Publication data available

∞ Printed on permanent acid-free paper, manufactured in accordance with the proposed ANSI/NISO Z 39.48-1992 and ANSI Z 39. 48-1984

CONTENTS

AUTHORS' ADDRESSES

P. Albuquerque, CEPLAC, Caixa Postal 1801, CEP 66000, Belém, Pará, Brazil
T. Andebrhan, CEPLAC, Caixa Postal 1801, CEP 66000, Belém, Pará, Brazil
J. Aragundi, c/o INIAP, Pichilingue, Apartado 24, Quevedo, Ecuador
F. Aranzazu H., ICA, Apartado Aereo 876, Manizales, Caldas, Colombia
R. Arias, INIAP, Pichilingue, Apartado 24, Quevedo, Ecuador
P. Buriticá C., Subgerente Investigación, Instituto Colombiano Agropecuário, ICA, Bogotá, Colombia
R.F. Colmenares, FONAIAP, Estacion Experimental, Bramon, Edo Táchira, Venezuela
J.C.B. Costa, CEPLAC, Caixa Postal 1801, CEP 66000, Belém, Pará, Brazil
F. James, Grenada Cocoa Association, Technical Division, Mt Horne, St Andrews, Grenada, West Indies
R.A. Lass, Cadbury Ltd, Bournville, Birmingham, B30 2LU, UK
G. Macias, INIAP, Pichilingue, Apartado 24, Quevedo, Ecuador
A.C. Maddison, 20a Moor Lane, Chessington, Surrey, KT9 1BW, UK
L.A. Maffia, Departamento de Fitopatologia, Universidade Federal de Viçosa, CEP 36570, Viçosa, Minas Gerais, Brazil
R.A.C. Miranda, CEPLAC, Caixa Postal 7, CEP 45600, Itabuna, Bahia, Brazil
S. Mohan, Cocoa Research Unit, University of the West Indies, St Augustine, Trinidad, West Indies
C. Moreira, INIAP, Pichilingue, Apartado 24, Quevedo, Ecuador
A.L. Pacheco, CEPLAC, Caixa Postal 1801, CEP 66000, Belém, Pará, Brazil
L.H. Purdy, IFAS, Plant Pathology Department, University of Florida, Gainesville, Florida 32611, USA
H. Reyes E, CENIAP, Apartado Postal 4653, Zona Postal 2101, Maracay, Estado Aragua, Venezuela
H.M. Rocha, CEPLAC, Caixa Postal 7, CEP 45600, Itabuna, Bahia, Brazil
S.A. Rudgard, CAB International, Development Services, Wallingford, Oxon, OX10 8DE, UK
R.A. Schmidt, Dept of Forestry, School of Forest Resources and Conservation IFAS, University of Florida, Gainesville, Florida 32611, USA
R.A. Setubal, CEPLAC, Caixa Postal 7, CEP 45600, Itabuna, Bahia, Brazil
R.B. Sgrillo, CEPLAC, Caixa Postal 7, CEP 45600, Itabuna, Bahia, Brazil
R. Silva-Acuña, FONAIAP, Estacion Experimental, Bramon, Edo Táchira, Venezuela
T.N. Sreenivasan, Cocoa Research Unit, University of the West Indies, St Augustine, Trinidad, West Indies
C. Suárez C., INIAP, Pichilingue, Apartado 24, Quevedo, Ecuador
B.E.J. Wheeler, Reader in Plant Pathology, Imperial College, Silwood Park, Ascot, Berks, SL5 7PY, UK
C. Williams, Grenada Cocoa Association, Technical Division, Mt Horne, St Andrews, Grenada, West Indies

EDITORS' PREFACE

The Monograph deals with the conception, planning, implementation, results and conclusions of the International Witches' Broom Project (IWBP), which was set up in 1985 with the aim of producing an economic management system for witches' broom disease of cocoa.

The contributions of the various sponsors, and the roles played by the participating organizations and scientists are described in the introductory chapter. Chapter 2 provides a review of what was, and what was not known from published literature about the cocoa–witches' broom pathosystem in 1989. The scope of the project and the approaches used are covered in Chapter 3, while Chapters 4 to 13 report on the field studies themselves in detail. The recent appearance of witches' broom in the important cocoa area of Bahia in Brazil is described in Chapter 14, before disease management recommendations are summarised and future prospects considered in the closing chapters.

The many man–years of field research in the IWBP in a total of six countries generated much useful information which was analyzed both in the individual countries and collectively. Even with a document of this size, certain information and analyses with less direct relevance to disease management had to be omitted. It is expected that more detailed treatments of certain aspects will emerge in scientific papers, and further analyses will be undertaken.

It is regrettable that the collapse in the world cocoa price has so severely reduced the financial returns from the crop, and it is hoped that this will not prevent further long–term research and evaluation of disease management recommendations.

The editors wish to thank Maureen Robinson, Alison Rudgard and Shirley Tchighianoff for typing and editing the manuscripts, and Annette Greathead for proof reading of the final draft. Stephen Rudgard also wishes to thank staff at CAB International Development Services for their tolerance and support during the production of the manuscript. Acknowledgements from Tony Lass and Hank Purdy to other organizations and individuals are to be found in Chapter 1, and the editors would like to endorse those acknowledgements.

The editors further wish to state that two words, phytosanitation and sanitation, have been used synonymously in this book to signify the removal of diseased parts from infected cocoa trees.

ABBREVIATIONS

a.s.l.	above sea level
BCCCA	Biscuit, Cake, Chocolate and Confectionery Alliance (UK)
°C	degrees Centigrade
CEPEC	Centro de Pesquisas do Cacau (Brazil)
CEPLAC	Comissão Executiva do Plano da Lavoura Cacaueira (Brazil)
CIDA	Canadian International Development Agency
cm	centimeter
CRU	Cocoa Research Unit (Trinidad)
DMMC	Data Management and Monograph Committee (of IWBP)
EET	Estación Experimental Tropical (Ecuador)
FONAIAP	Fondo Nacional De Investigacion Agropecuarias (Venezuela)
GCA	Grenada Cocoa Association
ha	hectare
HR	Horizontal Resistance
ICA	Instituto Colombiano Agropecuario (Colombia)
ICS	Imperial College Selection
IMC	Iquitos Mixed Calabacillo
INIAP	Instituto Nacional de Investigacion Agropecuarias (Ecuador)
IOCCC	International Office of Cocoa Chocolate and Sugar Confectionery
IWBP	International Witches' Broom Project
kg	kilogram
km	kilometer
ln	natural logarithm
LCT-EEN	London Cocoa Trade - Estación Experimental Napo
m	meter
mm	millimeter
P	Pound
PA	Parinari
PMC	Project Management Committee (of IWBP)
PSC	Project Scientific Committee (of IWBP)
SC	Selección Colombia
Sca	Scavina
Sil	Silecia
t	tonne
TSH	Trinidad Selected Hybrids
UF	United Fruit
VR	Vertical Resistance

SUMMARY
OF
DISEASE MANAGEMENT RECOMMENDATIONS

I. NEW PLANTINGS

Planting material

- Select resistant seed material, where it is available.

Fungicides

- In areas of very high disease incidence, protective compounds (copper–based) can be sprayed at 14 day intervals to protect seedlings during establishment.

Tree architecture and management

- Trees should be restricted to the first jourquette, by removing basal and canopy chupons;
- Maintenance pruning should keep canopy height at less than 3.5–4 m.

Phytosanitation

- All isolated plantings and others in low–risk areas: sanitation should be started in the 3rd year or as soon as disease appears, and practised conscientiously every year;
- Others in high risk areas: phytosanitation should be encouraged while it is cost–effective.

II. EXISTING PLANTATIONS

Decision to Implement Management Practices

The disease management practice that is most immediately applicable is phytosanitation. Before recommending this, the following factors need to be considered and a decision made regarding the possibility of the practice giving an economic return.

Current general practices in cocoa cultivation

Three management systems have been chosen from the range of possible options available:

- *Extensive systems* Plant trees: allow trees to develop naturally: collect ripe pods;
- *Minimal systems* As above with some additional cultural practices for tree management;

- *Intensive systems* As above with intensive tree management and some chemical inputs;

Only *Intensive* farmers are likely to adopt any labour- or cost-intensive practices designed to increase production.

Production and yield factors

- Pod loss to witches' broom can be too low to justify management measures (e.g. Grenada and Trinidad);
- The main factors affecting economics of broom removal are the rate of pay (per man day) (C) and the price (p) of cocoa per kg at harvest-time. Higher C/p ratios (i.e. > 3.0) are unfavourable.

Surrounding cocoa

- Where phytosanitation is practiced in neighbouring cocoa, pod losses are very likely to be reduced substantially by phytosanitation;
- Where phytosanitation is not practiced neighbouring cocoa:
  - *Discontinuous plantings* (farms separated by at least several hundred metres) the chances of successful reduction in pod losses are moderately strong;
  - *Continuous plantings* : significant reductions in pod losses are unlikely in high disease incidence areas, but are likely in low incidence areas.

Tree architecture and management

- Phytosanitation can only be implemented in well managed plantations with low canopies less than 3.5–4 m tall. In heavily infected plantations which are already much taller than 5 m, phytosanitation must be preceded by, or combined with, other practices, to rehabilitate the cocoa;
- Maintenance pruning should be completed frequently (as necessary) to remove basal chupons and canopy chupons;
- Broom removal may cause an increase of about 20% over unpruned plots in the total number of pods set independent of the proportion of pods lost to disease(s).

Techniques for Phytosanitation

If farm and economic conditions appear suitable, the following recommendations on the implementation of phytosanitation can be made:

Removal of brooms and other diseased parts

- The most frequently used tools are:
  - a hand-sharpened blade on a pole (with or without an extension)
  - secateurs on poles or long-handled pruners;
- Vegetative brooms should be cut off at least 15–20 cm below the point of infection;
- Diseased material on cushions should be thoroughly removed by cutting it off as close as possible to the bark;
- Diseased pods should be removed with their peduncles.

Efficiency of removal of diseased parts

- The removal needs to be as thorough as possible to have any chance of success.

Disposal of removed diseased parts

- All removed brooms and diseased pods should reach the ground and not remain suspended on branches within the canopy;
- Removal of diseased tissues from the plantation is recommended where:
  - annual sanitation has lapsed for more than one year;
  - basidiocarps are produced on brooms on the ground;
- Otherwise removed tissues can be left in the plantation, provided that:
  - removed diseased material is not heaped or left uncovered on the ground;
  - leaf litter is allowed to accumulate naturally, or placed over diseased materials;
- Broom removals can be done during periods of heavy leaf fall;
- Diseased beans should not be processed along with healthy beans.
- When large branches with brooms are removed from trees, the material should be cut up so that it is all in contact with the leaf litter to enhance decomposition;
- Entire trees may need to be eliminated if they are very heavily infected.

Other pod diseases

- Other pod diseases (e.g. *Moniliophthora roreri*) should also be managed if yield increases are to be secured.

Timing of phytosanitation

- Sanitation must be correctly timed;
- Sanitation must be scheduled outside the harvest periods of cocoa or other crops of local importance;
- *Primary removal* is to remove diseased material before the onset of the rains (to avoid basidiocarp production), and is best timed 1–2 months after the main period of broom formation is over;
- *Secondary removal* is to cope with the late–emerging brooms and those missed in the primary removal, and in most cases, to further reduce the number of productive brooms before the main period of pod set and development.

Regional guidelines for timing of brooms removals

Region	Primary	Secondary
Brazil – Amazon Region	August/September	October/December
Colombia (Llanos and Caldas[3])	February/March	June/July (none in Caldas)
Venezuela – Táchira	March/April	None
Venezuela – Barlovento[1]	March/April	September/October[2]
Grenada / Trinidad[1]	March/April	September/October[2]
Ecuador – Central Coastal Area[3]	July/August/September	November

[1] areas of high disease incidence only; [2] optional; [3] diseased pods to be removed every 7 days (for moniliophthora pod rot)

Chapter 1

INTRODUCTION TO THE INTERNATIONAL WITCHES' BROOM PROJECT

R.A. Lass & L.H. Purdy

WITCHES' BROOM DISEASE OF COCOA - AN OVERVIEW

Almost one-third of the world supply of cocoa is produced in the Americas but in many countries maximum production is severely curtailed by witches' broom disease. The pathogen (*Crinipellis perniciosa*) causes overdevelopment of cocoa meristematic tissues leading to hypertrophied and proliferated growth, to give the characteristic witches' broom (escoba de bruja/vassoura de bruxa/balai de sorcière). The fungus also infects many pods, which fail to ripen and the beans either become liquified or adhere to the pod husk and are commercially valueless. The disease is described in Chapter 2.

Witches' broom has long been known in the Upper Amazon region of Brazil and in Bolivia, Colombia, Ecuador, Grenada, Guyana, St Lucia, St Vincent, Panama, Peru, Surinam, Trinidad and Venezuela. In the Brazilian Amazon, pod losses have reached 95% on occasions, and almost all infected pods are a total economic loss. Several planting projects (e.g. Surinam, and Urabá, Colombia) have been abandoned because of the disease; in other areas, traditional farmers continue but with high disease incidence and low yields (e.g. Ecuador). With the colonisation of the Brazilian Amazon during the 1980s some 90,000 ha were planted with cocoa in a very short time, that figure would undoubtedly have been much higher had it not been for witches' broom. At the start of the International Witches' Broom Project (IWBP) in 1985, these countries together with the Amazon Region of Brazil produced some 198,300 t of cocoa which represented some 10% of world cocoa production. In 1992, the cocoa areas where the disease is present accounted for some 463,000 t of world cocoa production or nearly 21% (Anon, 1992). However, not all cocoa grown in these countries is affected by the disease and pod losses vary according to local conditions and local agronomic practices.

Much cocoa in South America is grown on holdings of between 15 and 30 ha, but there are plantings up to several hundred hectares in some of the more traditional cocoa regions. Plantations are normally scattered over large areas, and the crop seldom occurs as a monoculture over a whole region. There are over a million hectares in South America and the West Indies (Table 1.1), although at the start of IWBP at least half was not affected by witches' broom and produced about 65% of South American cocoa.

Disease Management in Cocoa: Comparative epidemiology of witches' broom. Edited by S.A. Rudgard, A.C. Maddison and T. Andebrhan. Published in 1993 by Chapman & Hall, London. ISBN 0 412 58190 6

TABLE 1.1
Cocoa production in 1985 for principal South American producers and estimated losses to witches' broom disease

Country	Production (tonnes)	Area[1] (ha)	Yield (kg/ha)	Growers receipts[2] ($ millon)	WB losses[3] ($ million)
Brazil[4]	67,500	90,000	750	89.6	38.4
Brazil[5]	325,000	550,000	590	507.0	-
Colombia	40,000	90,000	520	68.0	29.2
Ecuador	70,000	290,000	240	92.8	39.8
Trinidad	2,800	7,000	290	4.4	1.9
Venezuela	18,000	83,000	210	26.1	11.2
Total	522,500	1,110,000	-	787.9	120.5

[1] area of mature productive cocoa; [2] estimated using percentage of 1985 world price of $2,400/t; [3] estimated at 30%, therefore 30/70 of actual turnover; [4] Amazonian Region; [5] States of Bahia and Espirito Santo.

The disease was absent from the main producing region of Brazil in the states of Bahia and Espirito Santo in 1985 (Table 1.1; footnote 5). The escape appeared to be geographical but there was no reason to suppose that the exclusion would continue. On 24th May 1989, witches' broom disease was identified near Uruçuca in Bahia (discussed in Chapter 14). This was a new outbreak and was discovered some 2,000 km from the nearest known previous infection in the Amazon basin. CEPLAC[1] made a most strenuous attempt to eradicate infected material from that location (Chapter 14). Unfortunately, other outbreaks were discovered around Uruçuca in August and 125 km south of it in October of 1989. It was then obvious that the area involved was too large to make eradication a viable option, and it now seems most improbable that the disease will ever be eliminated from this major cocoa growing area.

COCOA PRODUCTION IN SOUTH AMERICA

Many factors affect the productivity of cocoa in South America, and it would be wrong to imply that witches' broom was the sole agent of importance. National average yields vary from 750 to 210 kg/ha (Table 1.1) due to a variety of factors. Unfortunately, there are few reliable estimates of losses to this disease over entire regions. In Colombia and Brazil, good yields are maintained even with high disease pressure. Average losses in these regions are probably about 30% (official sources), though up to 90% losses can occur where no cultural control is applied. Yields in Ecuador, Trinidad and Venezuela could probably be greatly improved with better disease management, although particularly in Ecuador another pod disease (*Moniliophthora roreri*) also causes severe losses.

The estimate of growers' receipts shown in Table 1.1 was calculated in 1985 at the

[1] Acronyms are explained at the beginning of the book.

start of the project, and was US $788 million. The grower receives between 55% and 65% of the world price. The world market price fell from $2,400/t in 1985 to $984/t in the autumn of 1992, and the same total recalculated at the latter date would be substantially lower, but still significant. If losses were fixed at an oversimplified estimate of 30% overall, this would mean that $120 million of potential earnings were lost annually in 1985, and $49.2 million at prices in the autumn of 1992. These global losses are set to increase as the disease continues to spread steadily through Bahia and Espirito Santo. There is also an unquantifiable, but nonetheless important, indirect loss of yield due to debilitation of the trees by the pathogen.

THE INTERNATIONAL WITCHES' BROOM PROJECT

The serious threat posed by witches' broom suggested that a co–ordinated, multi–centre, multi–national research programme on the disease should have high priority and so a project proposal was formulated in the early 1980s by scientists in countries where the disease was important. The International Office of Cocoa, Chocolate and Sugar Confectionery (IOCCC) with some other international donor agencies agreed in 1985 to fund a US $700,000 multinational research programme initially over a period of 5 years from 1st July 1985. IOCCC is the international trade association for the cocoa processing, chocolate and confectionery manufacturers. This body has had an enlightened attitude to supporting cocoa research for over three decades. The financial support for IWBP was supplied by 13 member associations [2].

The objective of the project was "to develop an economic management system to control witches' broom disease of cocoa in a number of cocoa growing environments". The project was extended to early 1993, to enable the completion of the data analysis and writing up of the results in this document. This Monograph records the scientific work of the International Witches' Broom Project in a way which it is hoped will prove useful to students of epidemiology, plant pathologists, cocoa scientists, and extension officers, as well as technicians faced with a similar plant pathology problem on other crops under other environmental conditions.

It is appropriate to record the very generous assistance to IWBP from the Governments of the participating countries in Latin America and the Caribbean where IWBP sites were established. Significant assistance in terms of facilities, transport and personnel were provided by the Governments of Brazil (CEPLAC), Colombia (ICA), Ecuador (INIAP), Grenada (GCA), Trinidad (CRU) and Venezuela (FONAIAP). We also wish to record our thanks to Cadbury Ltd, the University of Florida, USA, and Imperial College of Science

[2]The member associations of IOCCC were: Verband Der Susswarenindustrie, Austria; Association Belge des Chocolatiers, Belgium; Alliance of Danish Chocolate and Confectionery Industries, Denmark; Finnish Foods Industries Federation, Finland; Chambre Syndicale Nationale des Chocolatiers, France; Bundesverband Der Deutschen Susswarenindustrie, Germany; Associazione Industrie Dolciare Italiane, Italy; Nederlandse Cacao en Cacaoprodukten Vereniging, Netherlands; Norske Sjokoladefabrikkers Forening, Norway; Swedish Chocolate Confectionery and Biscuits Association, Sweden; Chocosuisse, Switzerland; Biscuit, Cake, Chocolate and Confectionery Alliance, UK; and Chocolate Manufacturers Association, USA.

Technology and Medicine, University of London, UK, for permitting the active participation of their staff in IWBP. The IOCCC Secretariat in Brussels facilitated the complex administrative arrangements of this multi-national, multi-lingual and multi-centred Project. Direct financial assistance from the Science and Technology for Development Fund (STD) of the European Community through Directorate General XII, from the UK Overseas Development Administration (ODA) and indirectly from the Canadian International Development Association (CIDA) is also greatly appreciated, and is acknowledged with gratitude. Without the assistance of all these agencies, the International Witches' Broom Project could not have existed.

OPERATIONAL STRUCTURE OF IWBP

It is felt helpful to describe briefly how the project was structured, organised and managed as this may prove useful to other scientists at a future date who might wish to embark on a similar enterprise. There was a substantial amount of preparatory work carried out in the late 1970s and early 1980s before a workable project proposal could be presented to potential donors for their consideration. It is important to note that without the seed money allocated by far sighted commercial donors to fund these initial discussions and the preparation of the many drafts of the project proposals which were required, the IWBP would never have started. The absence of such "seed money" could well be a factor which would prove limiting to any future projects. Major international donors are understandably reluctant to fund speculative gatherings of that nature to prepare project proposals. In particular the creative and financial support of the American Cocoa Research Institute (USA), the Biscuit, Cake, Chocolate and Confectionery Alliance, (UK), Cadbury Ltd, and the person of Hans Viskel (the President of IOCCC in 1984) contributed substantially to the start-up phase of IWBP.

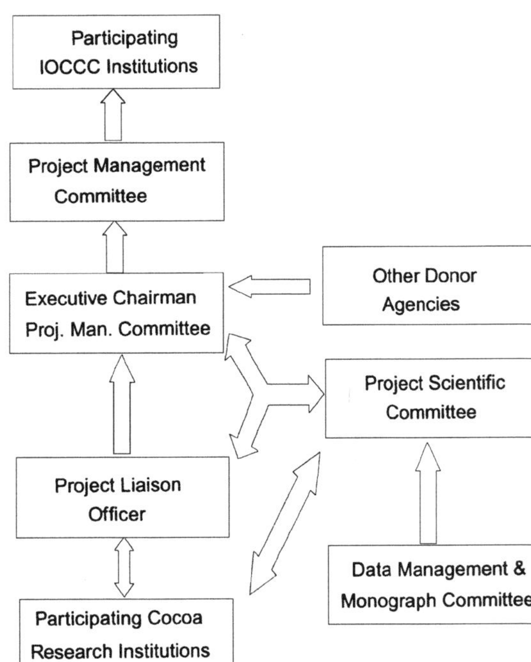

Figure 1.1. Organogram of the International Witches' Broom Project.

The organisational structure of IWBP is shown in Figure 1.1. This may appear complicated, in particular as IWBP had three committees - the

Project Management Committee (PMC), the Project Scientific Committee (PSC) and the Data Management and Monograph Committee (DMMC). However, the function and roles of these were clearly defined and agreed by everyone involved with the project right from the outset, and the arrangements worked well. The full terms of reference and the names of the members of these Committees of IWBP are listed in Appendix I. The PMC met regularly thoughout IWBP in locations around Europe, normally at yearly intervals. The PSC met during all the seven Workshops except the 2nd IWBP Workshop, and one additional meeting was held during the 10th International Cocoa Research Conference in Santo Domingo in May 1987. The DMMC was created in November 1988, and met seven times between then and April 1991.

From the very beginning of IWBP it became obvious that if the experimental programmes at the various sites were to be genuinely comparable, then it would be necessary for there to be regular co-ordinating visits in addition to having a series of regular workshops. The project was very fortunate in being able to recruit Dr S.A. Rudgard as the full-time project liaison officer. He had previously worked on this disease in the State of Rondônia, Brazil, and a regular programme of visits to the experimental sites was initiated right from the start of IWBP.

At the start of 1985, the Member Associations of IOCCC that were funding the IWBP invited R.A. Lass to manage the Project. He was duly appointed the Executive Chairman of IWBP, a post which he held until the end of the Project.

Dr L.H. Purdy kindly accepted the invitation to act as Chairman of the Project Scientific Committee throughout the duration of IWBP. In this role he also acted as Moderator at the seven IWBP Workshops held in various locations. Dr R.A. Schmidt accepted the invitation to act as Chairman of the Data Management and Monograph Committee. Dr B.E.J. Wheeler most generously agreed to undertake the task of acting as rapporteur at all of the IWBP Workshops, which he completed with great skill. His assistance in the production of the workshop reports is also gratefully acknowledged. Details of these Workshops are presented in Appendix II.

At the completion of the 7th IWBP Workshop an "Open Forum" on the present state of knowledge on witches' broom disease of cocoa was held (funded by IWBP) to which many individuals in Europe with research experience of the disease were invited. There were presentations by IWBP participants on their experimental conclusions and a stimulating and far-reaching debate ensued. A number of conclusions and recommendations of the Open Forum have been incorporated in this document.

FACILITIES AND PERSONNEL INVOLVED IN IWBP

The experimental programme of IWBP in each of the six collaborating countries depended on the national agricultural research organisation or a branch of the government education system. Facilities were provided by the co-ordinating organisation and/or university faculty in each country as previously named. It is obvious that without the generous and willing support of these government institutions it would not have been possible to conduct the experimental programmes. All those involved with IWBP wish to record their appreciation to all those bodies.

Five out of the total of 18 research sites were located within experimental plantings in research stations, two each in Colombia and Ecuador and one in Brazil. Otherwise they were located on commercial farms outside the direct control of the project staff. The excellent collaboration and support from these commercial growers is acknowledged with gratitude. Laboratories were not required for the research programme of IWBP. Office facilities and secretarial support were provided by the co-ordinating institution.

A standard set of meteorological equipment was provided by IWBP for each of the ten sites in the comparative epidemiology experiment. This consisted of the following: recording rain-gauge, a thermo-hygrograph, a surface wetness recorder, and a maximum/minimum thermometer. The Project also provided other major items as they were required by each country. Personal computers and some software packages were bought for the project co-ordinators in Brazil and Colombia. Two cars were purchased for the experimental programme in Brazil, and two motorbikes for the field staff in Ecuador. The considerable support to IWBP in Ecuador from ODA, in the form of A.C. Maddison's presence, vehicles and computing equipment is acknowledged with great appreciation.

Professional staff dedicated a proportion of their time to coordinating and supervising the IWBP field work and data analysis. The names of the research co-ordinators are listed below:

Brazil	:	T. Andebrhan
Colombia	:	F. Aranzazu H.
Ecuador	:	A.C. Maddison
Grenada	:	C. Williams
Trinidad	:	T.N. Sreenivasan
Venezuela	:	R. Acuna

Their assistance and enthusiasm contributed to the success of the project and is gratefully acknowledged. Research teams were composed either of professional technicians or trainees (see below), the majority of whom worked full-time on IWBP activities. Contract labour was required for the major operations in the phytosanitation studies, such as removal of brooms and pod harvests. This was funded by IWBP.

TRAINING ELEMENT OF IWBP

Two Brazilians operating the sites in Altamira and Tomé Açu both completed a Master's course on Plant Pathology in 1991 at the University of Viçosa, State of Minas Gerais, Brazil. Their theses were based on aspects of the epidemiology of witches' broom, and used the IWBP data. Professor L. Maffia of the Department of Plant Pathology, University of Viçosa, supervised their studies.

In Colombia, four site operators used the field experiments in the IWBP Comparative phytosanitation studies as thesis topics for their Ingeneiro Agronomo degrees. Two were based at the University of the Llanos at Vilavicencio, and the other two studied at the University of Caldas in Manizales, Colombia. Their degrees were awarded between 1988 and 1990. The external supervisor for all the students was Ing F. Aranzazu of the Instituto Colombiano de Agropecuaria.

Four students from Ecuador used the IWBP field experiments as the basis for their theses for the Ingeneiro Agronomo degree. Two studied the epidemiology of witches' broom, and the other two worked on aspects of phytosanitation of the disease, at one of the faculties at the Universities of Manabi, Guayaquil, or Cuenca, Ecuador. The external supervisor was Dr A.C. Maddison on attachment to the Instituto Nacional de Investigacion en Agropecuaria from the Overseas Development Administration, U.K.

The site operator for the comparative epidemiology experiment in Trinidad used the data collected during the IWBP field work as the basis for his MPhil thesis at the University of the West Indies. The work was supervised by Dr T.N. Sreenivasan of the Cocoa Research Unit, University of the West Indies, Trinidad, and the degree was awarded in 1990.

No staff undertook any training associated with IWBP in Grenada or in Venezuela. The full thesis titles and names of the students are listed in Appendix III.

THE EXPERIMENTAL PROGRAMME

The programme of experiments which has been completed during the life of the IWBP involved many more research centres and scientists than had been envisaged when the outline of the proposed experimental programme was drawn up in late 1984 and early 1985. This substantial additional experimental programme was possible because of the care and attention given by the research co-ordinators to the experimental layouts and to the disbursement of IWBP funds.

We believe that the IWBP is an outstanding example of international scientific co-operation and that it is a formula which might well be considered by anyone undertaking a similar programme of comparative research. It has been a privilege to be involved in this co-operative project, and we hope that in the future many will derive benefit and encouragement from reading this Monograph and that the studies will improve the situation of the many cocoa growers whose farms are currently devastated by this disease.

Chapter 2

THE PATHOSYSTEM

B.E.J. Wheeler & Carmen Suárez

THE PATHOSYSTEM IN OUTLINE

The cocoa tree (*Theobroma cacao* L.) occurs in the wild in tropical regions of South and Central America, particularly in the upper Amazon basin which is considered its primary centre of diversity. By the 16th century it was already well established as a crop in several areas within these regions and cocoa beans were used both as currency and as the basis of a drink, the 'chocolatl' of the Aztecs. Subsequently cocoa cultivation spread to other parts of South America and the Caribbean, to West Africa and to Malaysia and Indonesia. Virtually all the present world production of cocoa originates from these regions (Wood & Lass, 1985).

The cocoa tree itself is an intermittently-growing evergreen, exhibiting periods of leaf flushes alternating with periods of vegetative rest. In juvenile cocoa, leaf growth occurs at intervals of approximately 6–8 wk with leaves borne in a spiral. When young plants reach 1–2 m vertical growth ceases temporarily, up to five buds with short internodes emerge sideways from the end of the stem ('jorquette'), and on these 'fan' branches the leaves are alternate. Some months after formation of the first jorquette, buds lower down the stem grow out into shoots ('chupons') with an upright growth habit with spiral leaf arrangement, which grow vertically and eventually form new jourquettes above the first.

The flowers are cauliflorous, arising from 'cushions' on the bark, often in large numbers at certain times although in most regions a few flowers can be found throughout the year. The flowers are short-lived, and soon absciss unless pollinated by insects, of which midges are the most important. Not all of the very young pods ('cherelles') develop, many yellow, blacken and shrivel ('cherelle wilt'), leaving only a proportion of those set to mature, ripen and produce beans (Alvim, 1977; Wood & Lass, 1985).

Witches' broom disease derives its name from the groups of swollen shoots which develop from infections of vegetative buds and, sometimes, flower cushions. It was first reported from Surinam in 1895. Other outbreaks on cultivated cocoa were recorded in British Guiana (1906), Colombia (1917), Ecuador (1921), Trinidad (1928), Tobago (1939) and Grenada (1948). During this period it also became clear that the disease was endemic on wild species of *Theobroma* throughout the Amazon and Orinoco river systems (Baker & Holliday, 1957). In the last decade there have been serious outbreaks of witches' broom on new cocoa

Disease Management in Cocoa: Comparative epidemiology of witches' broom. Edited by S.A. Rudgard, A.C. Maddison and T. Andebrhan. Published in 1993 by Chapman & Hall, London.
ISBN 0 412 58190 6

plantings in the Amazon states of Brazil, particularly Rondonia, and most recently the disease has been reported from Bahia (1989), the most important cocoa-producing area in South America, from St Vincent (1988) and from Panama (1989). These new outbreaks raise again the question of its spread beyond South America (Wheeler, 1985*a*, 1987).

Although witches' broom was studied from the beginning of this century it was not until 1915 that Stahel showed convincingly it was caused by a basidiomycete fungus which he named *Marasmius perniciosus* (Stahel, 1915). Even so, because fruiting bodies (basidiocarps or mushrooms) of this fungus were not readily produced in pure culture, rigorous scientific proof of pathogenicity was not established until 1983 (Purdy *et al.*, 1983). The fungus was reclassified as *Crinipellis perniciosa* by Singer (1942), and this name is now used (Holliday, 1970). In the forests of the Amazon basin where it is endemic, *C. perniciosa* occurs not only on wild cocoa but on other related species of *Theobroma* and *Herrania*, on *Bixa orellana*, lianas and on wood from the understorey and litter (Thorold, 1975; Evans, 1978; Hedger, 1985; Bastos & Andebrhan, 1986). Isolates of the fungus within this complex differ in both morphology and pathogenicity. Three varieties, differing mainly in the colour and size of the basidiocarp cap (pileus), have been described by Pegler (1978) and there appear to be at least three groups of isolates with different pathogenicities (pathotypes), associated with cocoa, lianas and species of *Solanum* (Evans, 1978; Bastos & Evans, 1985; Hedger *et al.*, 1987). There is also considerable variation between isolates from cocoa growing in different regions (Wheeler & Mepsted, 1982, 1988; Andebrhan, 1988*a*; Bastos *et al.*, 1988; McGeary & Wheeler, 1988). Because of this pathogenic variability and lack of comprehensive cross-inoculation experiments there remains some uncertainty about the significance of isolates from other hosts in initiating disease on cultivated cocoa. Even isolates from other *Theobroma* species vary considerably in their ability to produce brooms on cocoa (Fonseca *et al.*, 1985).

On cocoa, basidiospores of *C. perniciosa* infect actively-growing (meristematic) tissue and induce a range of symptoms on vegetative shoots, flower cushions, flowers and pods. The diversity of these symptoms reflects the type of tissue involved and its stage of development, the genetic constitution of the host and the pathogenicity of the fungal isolate, plus effects of environment on one or more of these factors. Understanding the nature of these symptoms has been an integral part of research on witches' broom, including the investigations described in this monograph for which descriptions of symptoms were produced based on earlier accounts and practical experience (Rudgard, 1989). The described symptoms cover infections on the elongating vegetative shoots (necrotic leaf, stem swelling, axillary and terminal brooms), on flowers (single, simple and compound flower brooms) which sometimes result in distorted fruit (strawberry fruit or chirimoya) or a vegetative cushion broom, and infections of cherelles and pods.

The destructive effects of witches' broom are scarcely in doubt. Although other factors were involved, its role in the decline in cocoa production in Surinam, Trinidad and Ecuador are well documented (Baker & Holliday, 1957). Accurate estimates of losses are less easy to obtain mainly because the fungus reduces pod yield not only directly but also indirectly through its debilitating effects on tree growth. However, even in new plantings in Rondonia direct loss of pods reached 50% by the sixth year (Rudgard, 1986*a*), and losses over 90% have been reported (Evans, 1981). Annual losses in Brazil, Colombia, Ecuador, Trinidad and Venezuela have collectively been put at $120 million or about 30% of expected production

(Rudgard & Lass, 1985).

ASPECTS OF THE PATHOSYSTEM

Disease Cycle

There are two phases in the disease cycle (Figure 2.1). First *C. perniciosa* invades young tissue, induces hypertrophy and hyperplasy and lives as an intercellular obligate parasite (biotroph). This phase ends as the hypertrophied tissue dies. The fungus then invades cells, grows as a saprotroph and eventually, when conditions are favourable, produces basidiocarps in which the infective basidiospores form (Evans, 1981; Hedger, 1985; Wheeler, 1985*b*).

Infection of young cocoa tissue begins when germ–tubes from basidiospores enter through stomata or penetrate directly through the epidermis or trichomes (Stahel, 1919; Cronshaw & Evans, 1978; Frias, 1987; Sreenivasan & Dabydeen, 1989). After stomatal entry, hyphae from substomatal vesicles colonize the host tissues intercellularly (Frias, 1987), but the exact sequence of events following direct penetration has not been established. The time taken for symptoms to appear can vary considerably (3–14 wk), but is usually about 5–6 wk (Baker & Crowdy, 1943). The fungus apparently causes an hormonal imbalance, in ways not yet understood, so that host cells are larger than usual, particularly those of the cortex and pith (Orchard & Hardwick, 1988; Orchard *et al.*, 1989), and the tissues become swollen. On vegetative shoots, apical dominance is lost, axillary buds develop, and a broom forms.

In green brooms the mycelium of the fungus is composed of relatively wide (5–15 μm), intercellular hyphae, often appearing swollen and flexuous but without clamp connections (Stahel, 1915; Pegus, 1972; Delgado, 1974; Delgado & Cook, 1976; Suárez, 1977; Calle, 1978; Evans, 1980; Calle *et al.*, 1982; Aragundi, 1982*a*; Danquah, 1986; Frias, 1987). This mycelium colonizes the various parenchymatous tissues of the broom but to different extents. Its cells contain one nucleus and it is considered to be monokaryotic (Pegus, 1972; Evans, 1980; Mayorga, 1989). It has not yet been grown on artificial media but mycelium of similar appearance has been obtained by inoculating cocoa callus with basidiospores (Evans, 1980; Fonseca, 1988).

Brooms usually remain green for only a relatively short period (Aranzazu, 1982*a*; Cifuentes *et al.*, 1982*a*). They begin to dry out from the shoot tips, turn brown in about 5–6 weeks and then progressively become dry. There are changes in the fungal mycelium associated with the death of the broom tissue although the relationship between these events is still not entirely clear. Fungal hyphae are to be found inside cells and the broom becomes extensively colonized by a mycelium with binucleate cells which appears to develop following anastomosis between adjacent cells of uninucleate hyphae. As the broom dries so these intercellular hyphae become divided into thick–walled chlamydospores (see report of T.N. Sreenivasan in Lass & Rudgard, 1987; Mayorga, 1989).

Basidiocarps do not form on brooms which have recently dried; there is usually a dormant (induction) period of at least 4 months before they are produced (Baker & Crowdy, 1943; Aranzazu, 1982*a*; Cifuentes *et al.*, 1982*a*), even if the brooms are transferred to favourable regimes of wetting and drying (Rocha & Wheeler, 1985).

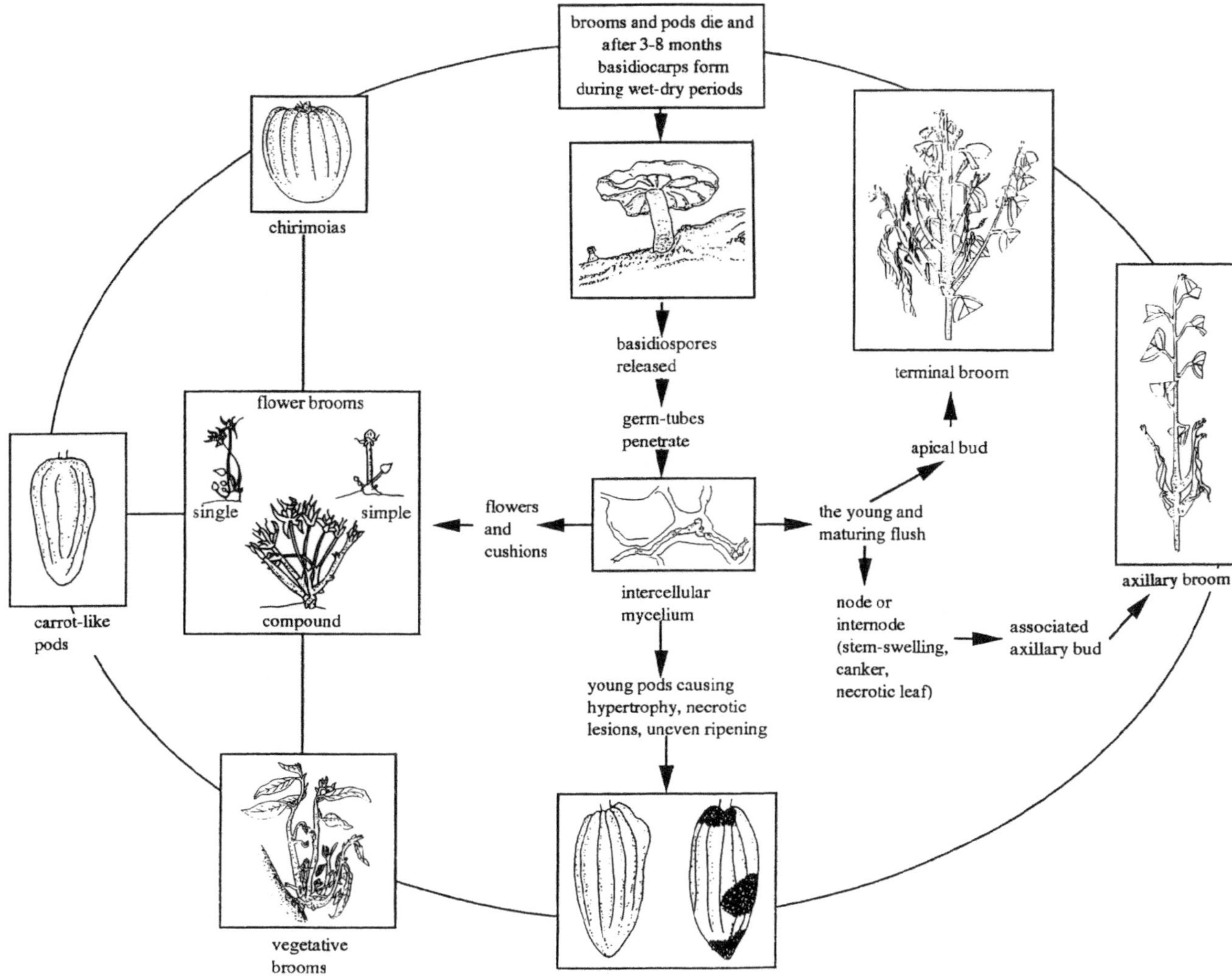

Fig. 2.1 Disease cycle of witches' broom on cocoa caused by *Crinipellis perniciosa*

Chlamydospores germinate in brooms kept in these conditions and give rise to hyphae, 1.5–3 μm wide, with clamp connections and binucleate cells (Pegus, 1972; Mayorga, 1989), which grow through all the tissues and eventually aggregate beneath the bark and form basidiocarps. This saprotrophic mycelium (Stahel, 1915), can be readily isolated on agar media and produces degradative enzymes *in vitro* (Lindeberg & Molin, 1949; Krupasagar & Sequeira, 1969), but it cannot infect cocoa.

Two aspects of this disease cycle have attracted most research – the source and production of inoculum, and disease development.

Source and Production of Inoculum

In his original publication on witches' broom, Stahel (1915) noted that basidiocarps formed on brooms, infected flower cushions and pods only during periods of rain. Later, he found that at these times basidiocarps could be produced on detached brooms laid on slatted tables beneath the shade of a *Ficus* sp. tree, some brooms remaining productive for 1½ years. Alternate wetting and drying of the brooms appeared beneficial; persistence in depressed basidiocarp production (Stahel, 1919). Since then there have been several detailed investigations of fruiting on brooms on trees (Baker & Crowdy, 1943), and on detached brooms kept in the field (Aranzazu, 1982*a*, *b*; Rudgard, 1986*b*), or in cabinets in the laboratory (Rocha, 1983; Rocha & Wheeler, 1982, 1985). These studies indicated influences of both host and environment.

Generally, more basidiocarps were produced on vegetative and cushion brooms than on pods and numbers increased with broom size (Baker & Crowdy, 1943). On brooms kept in cabinets, more basidiocarps formed at nodes than on internodes but productivity also varied on brooms of similar size taken from different clones (Rocha & Wheeler, 1985).

The wetting and drying of brooms appear to be important environmental factors. In cabinets, most basidiocarps formed with a wet period of 8 h/day, progressively less with periods of 16 h and 23 h. No basidiocarps formed with only one hour of wetness per day. Also, in an 8 h wet/16 h dry regime, more basidiocarps were produced at 20–25°C than at 25–30°C and none formed at 30–35°C (Rocha & Wheeler, 1985). Comparable results were obtained in the field by Rudgard (1986*b*). Brooms were most productive with moderate amounts (8–16 h) of wetness per day and less than 4 h or more than 20 h were inhibitory. No fruiting occurred at mean air temperatures over 30°C or below 20°C. In this field situation the induction of fruiting required a total of at least 17 rainy days but brooms in their second year of production responded more rapidly than younger brooms. Fruiting was reduced on brooms on the leaf litter, probably because these remained wet on average 6 h/day longer than brooms hung in the canopy.

Two of these investigations indicated clearly the average number of basidiocarps formed per broom because all basidiocarps were removed when they were counted. The data of Baker & Crowdy (1943) are most extensive and particularly useful since they were obtained in the field. Productivity over 9 months on terminal and lateral fan brooms of different size varied from about 5 to 39 basidiocarps per broom. In the same period small cushion brooms (mainly star blooms) produced on average three basidiocarps and large, mainly vegetative, cushion brooms 89 basidiocarps. In experiments reported by Rocha

(1983), vegetative brooms hung in cabinets with the most favourable regime (8 h wet/16 h dry) produced on average 10 basidiocarps in 5 months.

Patterns of fruiting were also apparent in these studies, brooms producing flushes of basidiocarps even when kept in standard regimes. These flushes may be associated with successive cycles of absorption of nutrients into the mycelium and their depletion by fruiting. The patterns varied with source of brooms (Rocha & Wheeler, 1982), and even within brooms from one locality (Rudgard, 1986*b*). However, it is the seasonal pattern of fruiting in any one region which is of most practical interest since this indicates when inoculum is available. These patterns clearly indicate a general association of basidiocarps with rain (Baker & Holliday, 1957; Aranzazu, 1982*a*; Cifuentes *et al.*, 1982*a*; Andebrhan, 1985*a*). Attempts to correlate numbers of basidiocarps with weekly amounts of rain have not been successful (e.g. see Baker & Holliday, 1957), but Andebrhan (1985*a*) obtained significant correlations between basidiocarp production and numbers of days with rain and evaporation (real) less than 2 mm in the same months at sites within the Brazilian Amazon.

There have been fewer studies on basidiospore release and dispersal. Stahel (1915) first showed that basidiospores were discharged at night and he linked this with temperature changes. These findings were confirmed and expanded by Baker & Crowdy (1943) in Trinidad. They found that basidiospores were released between 18.00 and 06.00 h with most basidiocarps releasing spores between 22.00 and 04.00 h, provided the humidity was high enough to keep basidiocarps turgid. The most extensive data are provided by Solorzano (1977), based on hourly counts of basidiospores caught in a Hirst trap near a source of brooms, from March to July 1976 at Pichilingue (Ecuador). Overall, maximum release occurred between 22.00 and 04.00 h, but some basidiospores were caught from 16.00 to 11.00 h the next day. Basidiospore release began when relative humidity rose to > 95% and temperature fell below 24°C. During these periods fewer spores were caught when it rained, presumably because they were washed out of the air.

The effects of temperature and humidity have been studied more extensively in controlled environments (Baker & Crowdy, 1943; Bastos, 1982; Rocha & Wheeler, 1985). The limits for spore release appear to be 10°C to 30°C and 80% r.h. to near saturation, but discharge of basidiospores is optimal at 20–25°C and 80% r.h.

These factors also affect longevity and spore productivity of basidiocarps. Experiments in Trinidad with brooms kept in a moist atmosphere indicated a decline in the period over which basidiocarps produced spores from 85 h at 60°F (15.6°C) to 70 h at 68°F (20°C) and 35 h at 80°F (26.7°C) (Baker & Crowdy, 1943). Earlier observations by Stahel (1919), based on 20 basidiocarps, also kept in moist chambers (but temperature unspecified), similarly showed that most of them released spores over 2–4 days, with the larger basidiocarps generally lasting longer.

In later work, Khonje (1984), estimated the numbers of spores released by basidiocarps on brooms hung in cabinets at 20–25°C with a daily regime of 13 h wet/11 h dry. Three basidiocarps produced spores over 4, 6 and 8 days, the maximum release was 45– , 105– and 70–thousand basidiospores/h, giving a total production of about 2.0, 1.9 and 3.5 million spores respectively. Data presented by Rocha & Wheeler (1985) similarly indicated mean productivities, in thousands of basidiospores/h, of 43–119 at 20°C and 75–146 at 25°C with relative humidities ranging from 80% to near saturation.

There are no comparable data on basidiospore production in the field but in Rondonia basidiocarps remained fully expanded and turgid for up to 5 days and traps at various heights beneath the cocoa canopy gave estimates between 1,500 and 4,000 basidiospores/m^3 air (Rudgard, 1987). In this situation most spores sedimented out slowly but during rain basidiospores may be distributed in water running down branches and the trunk of the tree and thus serve as inoculum for pod infection. The potential of this inoculum was demonstrated by Andebrhan (1988*b*) by placing germinating cocoa seeds in traps collecting run-off water which were protected from airborne spores. About 20% of the seedlings subsequently developed witches' broom symptoms. The possibility of long-distance spread (50–70 km) of basidiospores is indicated by work in Ecuador using seedlings as trap plants (Evans & Solorzano, 1982). Also, dispersal and infection gradients have been established up to 285 m from an inoculum source using rotorod spore traps and susceptible cocoa seedlings (Aragundi *et al.*, 1988).

Disease Development

Clearly, *C. perniciosa* has a considerable capacity for producing inoculum but there are few data which indicate inoculum efficiency, though Andebrhan & Almeida (1987) showed that percentage infection was reduced with decreasing inoculum dose. Inoculum efficiency is certainly affected by the nature of the cocoa material, as shown in experiments with standard inocula. Frias (1987) sprayed young seedlings to give maximum doses on flush growth of 10–, 30– and 60–thousand spores. At the lowest inoculum dose there was over 75% infection of Catongo but less than 50% on Scavina 6 and Scavina 12. At higher inoculum levels this difference was less apparent. Similar results were obtained by Mayorga (1989) who used measured inocula on agar blocks to inoculate seedling hypocotyls. Inoculations of ICS 39 x Sca 9, PA 46 x IMC 67 and TSH 792 x IMC 67 with 1,000 spores gave 100%, 45% and no infected seedlings respectively.

The age of the cocoa tissue at inoculation also influences infection. Young flushes, particularly the developing buds are most susceptible. As the flush develops it becomes more resistant, and inoculations of shoots with hardened leaves and a dormant bud are usually unsuccessful (Baker & Crowdy, 1943). However, inoculation of a hardened flush immediately beneath the growing tip can cause a slight swelling leading to the formation of a canker (Cronshaw & Evans, 1978). The susceptibility of pods also changes with time. The exact period of susceptibility is difficult to determine because symptom expression (necrosis) is delayed, but it appears to be about 12 wk from pod set, cherelles over 6 cm long being seldom infected (Baker & McKee, 1943; Baker & Crowdy, 1944; Andebrhan, 1985*a*; Rudgard & Butler, 1987).

Infection is also influenced by environmental factors of which the most important appears to be water films on susceptible tissues. Basidiospores of *C. perniciosa* germinate rapidly in water, germination occurring within 3–4 h at 22–24°C (Baker & Crowdy, 1943). Under these conditions the minimum period for infection may also be as little as 4 h (Khonje, 1984).

Disease development is thus basically related to the frequency of wetness periods and the available amounts of both fungal inoculum and susceptible host tissue when these periods occur. However, prior to the IWBP, few studies provided such data for describing and analysing witches' broom epidemics. The main aspects responsible for these shortcomings, some of which remain, are that host surface wetness is technically difficult to monitor, amounts of inoculum are more readily determined indirectly using numbers of basidiocarps than directly from counts of trapped basidiospores, the diversity of symptoms and range of incubation periods complicate disease assessment, and estimating the abundance of infection courts has daunted most investigators.

Nevertheless, some analyses were attempted and relationships established. In Trinidad, Baker & Crowdy (1944) found a positive correlation between numbers of basidiocarps and the total number of brooms produced 5 wk later. A similar correlation was obtained by Aranzazu (1982*b*) in Colombia but using a 2 month interval. Positive relationships between flowering and cushion infections have also been established for data from Trinidad (Baker & Crowdy, 1943), and Rondonia, Brazil, where the incidence of pod disease was also related to the number of wetness periods longer than 4 h and the numbers of basidiocarps 12 wk earlier (Rudgard & Butler, 1987).

It has proved more difficult to discern significant relationships between flushing and infection; correlations are low although flushing obviously has some effects on disease incidence (Baker & Crowdy, 1944). Similarly, the relationship between infection and shade, which itself influences flushing, is not clear. In the Brazilian Amazon, Andebrhan (1985) observed paired trials of shaded and unshaded cocoa in two locations; significantly more vegetative brooms were found per tree in the unshaded plot in Altamira where the cocoa was not pruned clear of brooms, whereas there was more infection in the shaded site in Tomé Açu which had regular sanitation.

MANAGING THE PATHOSYSTEM

The purpose of examining relationships between inoculum, host growth, environment and disease is to plan disease management strategies more effectively. The methods fall mainly into three broad groups, sanitation, spraying with chemicals, and the use of resistance, but in operation they often entail modifications in the way cocoa is grown. Each method or combination of methods needs to be appraised in terms of cost and benefit and against a background of fluctuating cocoa prices this presents other problems (Rudgard & Andebrhan, 1988; Laker & Rudgard, 1989).

Sanitation is based on the concept that by removing infected parts the production of inoculum will cease or at least be reduced to low levels depending on the efficiency of the removal. It has remained the basis of witches' broom control since it was first proposed by Ritzema Bos at the beginning of this century (Stahel, 1915; Thorold, 1975). In practice, sanitation entails pruning of brooms since diseased pods generally contribute little inoculum and can be removed during harvesting. The frequency of pruning is related to broom formation, basidiocarp production and cropping period in the particular area and takes advantage of any dry season when flushing is low and dry brooms can be seen more readily. Essentially it depends on the long (induction) period for basidiocarps to develop on brooms which means that the latent period can be several months. There is generally only one disease cycle per rainy season, that is, within the span of one year the disease is monocyclic.

In summarising data from Trinidad, Baker & Holliday (1957) recommended two prunings, in late April and in October. The first pruning mostly removed the many brooms formed from January to March. The second pruning dealt with any material which was missed and any new brooms formed from July onwards before they produced basidiocarps. It also minimised infection of pods in the main harvest. Two prunings are also advised in the Brazilian Amazon (Andebrhan, 1985*a*). In Rondonia the recommendation is one main pruning in the dry season (June–August), which must be completed before the first large wet-season flush, followed by a second pruning before mid-December (Rudgard, 1987). Whether the excised brooms can be left on the ground or should be removed has often been disputed and focusses attention on the continuing ability of these brooms to form basidiocarps. In some situations, especially where the foliage canopy is intact, fallen brooms remain wet for long periods and are unproductive. Particularly, brooms incorporated into wet leaf litter soon fail to produce basidiocarps, they become colonized by other fungi and the mycelium of *C. perniciosa* survives only for a few months (Hedger, 1985, 1988). In other situations wetting and drying of brooms on the ground facilitates basidiocarp production and here excised brooms should be covered with leaf litter (or banana leaves) or removed from the plantation (Rudgard, 1987).

In the earliest work on witches' broom, an extreme sanitation method was tried, that of pollarding, removing the entire crown of the tree, leaving only the trunk and part of the main branches (Stahel, 1915). It was not widely adopted but has been used again recently in Colombia to rehabilitate old cocoa and in Pará state, Brazil, apparently with success (Andebrhan & Almeida, 1985; Grisales & Cubillos, 1985).

The value of more orthodox sanitation has been questioned, especially following disappointing results as in large-scale experiments in Trinidad during the 1940s (Baker & Crowdy, 1944). Several factors can militate against the success of sanitation. Brooms are difficult to remove from tall trees and it is relatively easy to miss some infected material especially with cocoa of high susceptibility where numbers of brooms are high (Bastos & Evans, 1979). Also, as with all sanitation programmes, there is a need to start early in the epidemic and it is labour-intensive (Almeida & Andebrhan, 1988). However, two basic questions remain: how effective is sanitation, and what risk is there subsequently of re-infection by basidiospores from a distant source? Information is sparse on both aspects. Removal of brooms must be efficient if sanitation is to work. A model, based on data from Rondonia, suggested that about 95% removal of brooms was necessary to reduce pod loss through witches' broom by 50% (Rudgard & Butler, 1987). Nevertheless, beneficial effects have been demonstrated such as delaying the onset of the witches' broom epidemic and reducing its infection rate (Andebrhan, 1985*a*), reducing the incidence of brooms (Rodriguez, 1988), and increasing yields (Almeida & Andebrhan, 1988). Data on basidiospore dispersal from Ecuador (Evans & Solorzano, 1982; Aragundi *et al.*, 1988), clearly indicate the potential for spread between plantations, although the experiments were in open areas and do not allow entirely for filtering-out of spores by trees. While some incoming inoculum could re-establish infections in a plantation where brooms have been removed, inoculum generated within the plantation is likely to be more important (Andebrhan, 1988*b*), and continuing phytosanitation should deal with this.

As with sanitation, the use of chemicals to control witches' broom dates from the early years of this century when van Hall and Drost recommended sprays of Bordeaux mixture (Stahel, 1915), but spraying has seldom given sufficient control to offset its costs (Baker & Holliday, 1957). However, the introduction of systemic fungicides has renewed interest in the chemical control of witches' broom and many materials have been screened in both laboratory and field. The salient features of this research are reviewed by Laker & Rudgard (1989). Two strategies now seem possible. One is to protect the pods using a broad-spectrum residual fungicide, such as copper; the other is to spray the canopy with a systemic fungicide both to inhibit broom formation and to eradicate mycelium from shoots and flowers. There are application problems to be resolved with both types of chemicals. Deposits on pods are reduced by rain and host growth, and spraying the canopy is a relatively inefficient way of placing fungicides near infection courts. The alternative of applying to the soil a chemical which then readily translocates in the cocoa tree raises problems of residues in the beans and/or contamination of water sources through leaching from soil.

Recently there has been some interest in 'biological control', using that term in a wide sense. Within cocoa plantations there is a natural biological control of *C. perniciosa* in fallen brooms by other micro-organisms some of which might be exploited either by manipulating the environment of the brooms or by using the 'antibiotic' substances they produce (Bravo & Hedger, 1988). The discovery of a hyperparasitic fungus, *Cladobotryum amazonense*, on basidiocarps of *C. perniciosa* has also prompted research on its potential in 'biological' control (Bastos, 1984; Bastos *et al.*, 1981, 1986).

The possibility of selecting for resistance was considered in the earliest reports of witches' broom. Stahel (1915) noted that not all trees in Surinam were equally susceptible, but because he considered the disease could be controlled he advised the selection of the best bearing trees rather than those which appeared less susceptible. This approach was also used in Trinidad after extensive screening within the Trinitario population failed to detect any highly resistant material. However, in this further programme using yielding ability as the main criterion in a situation of high pathogen pressure, there was also a bias towards witches' broom resistance and some clones, such as ICS 95 showed considerable resistance (Baker & Holliday, 1957; Thorold, 1975). Resistance here was expressed as fewer developed brooms (Baker & Crowdy, 1944), a feature subsequently noted elsewhere (Cifuentes *et al.*, 1982*b*).

Meanwhile, a search for immune or resistant trees was made in South America in several expeditions, notably those of F.J. Pound (Toxopeus, 1981). Three clones from Pound's collection designated Sca 6, Sca 12 and IMC 67, particularly contributed a high degree of witches' broom resistance to progeny bred from them to improve bean size. These Trinidad Selected Hybrids (TSH) have been widely planted in Trinidad since the late 1950s and appear to have been a major factor in reducing witches' broom there to its present relatively low levels (Laker *et al.*, 1988*a, b*). However, when apparently resistant selections of Sca 6 were sent to Ecuador they became severely infected, probably because they encountered a more aggressive strain of *C. perniciosa* (Wheeler & Mepsted, 1982, 1988).

Breeding for witches' broom resistance was subsequently started in other countries, often following further searches for germplasm, for example, the programme at the Tropical Experiment Station, Pichilingue, with material collected by Desrosiers and von Buchwald (Thorold, 1975; Soria, 1970; Allen & Lass, 1983; Allen, 1987). The main features of resistance breeding to the early 1970s are reviewed by Bartley (1977). While there are selections with some resistance to isolates of *C. perniciosa* in Brazil, Trinidad and Venezuela, the level of resistance in commercial cocoa is still less than satisfactory. In other countries such as Bolivia, Colombia and Ecuador which have more virulent isolates the situation is even less good.

The practical problem of screening cocoa germplasm collections for resistance to *C. perniciosa* remains. New, rapid methods are being sought such as the use of cocoa callus (Fonseca & Wheeler, 1990) or culture filtrates of the fungus (Muse *et al.*, 1988), as yet with little success. A bulk screening method was tested by Frias (1987) as mentioned before, and is now being used for further experiments. Evaluation of symptoms remains a problem in all types of resistance tests.

There are other aspects of the pathosystem which particularly require a greater understanding. They are the interactions between weather, inoculum and host phenology. Only if these interactions are properly understood can sanitation and chemicals be used rationally in a management programme. The acquisition of appropriate data to achieve this understanding is the subject of the chapters which follow.

Chapter 3

THE IWBP STUDIES

S.A. Rudgard, T. Andebrhan, A.C. Maddison & R.A. Schmidt

INTRODUCTION

The objective of the IWBP was to develop practical and reliable economic methods of managing the disease in a number of cocoa growing environments. The review of the pathosystem in Chapter 2 showed that a large amount of research had already been completed before the IWBP was conceived. However, various aspects of the epidemiology of the disease remained poorly understood, meaning that management of the disease was not always based on an adequate knowledge of the pathosystem.

It was known that witches' broom incidence varied considerably across its distribution, and disease management had been addressed on a regional basis. In some areas where incidence was moderate, practices had been developed to reduce pod loss, the economic effect of the disease, but these were often too expensive to apply when prices were low. In other areas, sustainable or effective management systems had not been developed, and high disease incidence had forced growers in some countries or regions to abandon cultivation of cocoa,as described in Chapter 1.

Existing data on the witches' broom pathosystem provided significant information on the disease cycle and the epidemic. However, in relatively few investigations had data been collected concurrently on sporophore production, host growth, disease incidence and meteorological conditions. Usually data were available only on one or two variables. A wider and deeper knowledge of the complex pathosystem based on comparisons between different regions was a necessary step towards improving disease management.

To achieve this a research programme was devised which would examine systematically infection dynamics of witches' broom relative to time (both seasons and years) and space. Scientists involved in the project debated the character and form of the experimental programme throughout the first years of the project. Differences in opinion existed on the most appropriate methods by which to achieve the required understanding of the pathosystem, and the experimental plans were chosen after considering the practical needs of the cocoa industry, aspects of theoretical plant pathology, and the financial constraints of the project budget. The experimental programme consisted of three parts, namely one comprehensive study of disease dynamics, and two studies of disease spread. However, it is

Disease Management in Cocoa: Comparative epidemiology of witches' broom. Edited by S.A. Rudgard, A.C. Maddison and T. Andebrhan. Published in 1993 by Chapman & Hall, London. ISBN 0 412 58190 6

important to note that much research on witches' broom that was outside the scope of the project was completed in parallel to it.

COMPARATIVE EPIDEMIOLOGY EXPERIMENT

This experiment was intended to study the phenology of infection courts, sporophore production and disease in cocoa plantations that were representative of several regions. Scientists at the 1st IWBP Workshop developed a standard design for the experimental methods in the form of IWBP's *Protocol 1* (Appendix IV), aimed at providing answers to three basic epidemiological questions:

- Which diseased tissues in which locations in the plantation are important in producing basidiocarps, and when do they produce them?
- What susceptible tissues are available for infection during the year, what is their relative abundance, and what proportion becomes infected?
- Which meteorological factors influence the production of inoculum and the infection of cocoa?

Nine sites were finally established in areas where the disease was present, and one site in a region without witches' broom (Bahia, Brazil). Sites differed in host material, layout and other features. Despite attempts to standardize methods, the protocol was not always followed exactly, but almost all of the differences were minor. Sources of variability are detailed in Chapters 5–10 which describe the sites and present results country by country, and in the comparative analysis in Chapter 11.

The sites were to be cleared of inoculum sources at the start of the experiment, and a known number of sources were arranged in specified positions. The sites differed in their degree of isolation and the amount of disease present before the study, as is shown in Chapter 4. The intention was that inoculum from within the site should be primarily derived from the arranged sources. Background inoculum could not be excluded. Infection was allowed to progress for two years with no further interventions, and most important aspects of disease, host and environment were recorded.

The epidemics at the individual sites were analysed (Chapters 5–10), and then a comparative analysis was made of all the sites (Chapter 11). Certain features of the witches' broom–cocoa pathosystem restricted the scope of these analyses. For example, to examine the interaction of inoculum, host tissues and climate, it is necessary to know when infection occurs and not only when symptoms appear. Incubation periods were known to fluctuate widely (Chapter 2), and precise identification of the week in which infection occurred was not possible. However, some analyses were made using reasonable estimates of incubation periods to allow for delays in appearance of disease symptoms.

Comparisons amongst sites examined climate, the timing and relative rates of sporophore production (not spores), the availability of host tissue, symptom types and timing, and disease indices relating numbers of infections to numbers of infection courts. As the study covered just two years, and therefore only two cycles of the disease, disease progress curves and infection rates (Vanderplank, 1963) could not be used to compare the epidemic among locations. The type of comparative epidemiology described by Kranz (1974) for polycyclic diseases was not appropriate.

GRADIENTS FROM ISOLATED SOURCES

Disease gradients are important aspects of witches' broom epidemiology that have seldom been studied. The 1st IWBP Workshop decided that the Project would "elaborate plans for disease gradient experiments", although this was not actually done until the 3rd IWBP Workshop. The group acknowledged that there is no single disease or inoculum gradient for witches' broom, owing to variations in many factors such as shape, strength and location of the inoculum source, climatic factors and topographic features.

The second study was designed to quantify the numbers of spores (spore gradients) and/or disease (disease gradients) as a function of distance from an isolated discrete source of inoculum of known size. It was intended that the sub-programme would consist of a series of experiments, however only one site was established, and only disease gradients were examined there. A quantity of sporulating brooms acting as inoculum sources was placed at one end of a forest clearing, and disease gradients were detected using susceptible cocoa seedlings as trap plants located at several distances from the source.

COMPARATIVE PHYTOSANITATION EXPERIMENTS : GRADIENTS IN PLANTATIONS

The third study was intended to examine the spread of witches' broom within a mature cocoa plantation. In contrast to disease gradients from isolated sources studied using cocoa seedlings, there was no published information regarding gradients of infection within a plot of mature cocoa trees. An experiment was designed at the 4th IWBP Workshop to study the effect of an infected area on disease in shoots, flower cushions and pods in an adjacent area cleared of infectious material. This had the benefit of combining a study of within-farm spread of disease with a study of the efficacy of phytosanitation, and providing information of inter-plot interference. It was hoped that the improved knowledge would help to explain why phytosanitation was successful in certain situations but not in others, and to provide a sound basis for designing experiments and demonstration plots. A total of six sites was established in four of the six countries following a standard design in the form of the IWBP's *Protocol 2* (Appendix IV).

Phytosanitation could have been considered separately from within-plot spread. One alternative was to consider complete sanitation of an area of cocoa, which would have been a straightforward test of whether the disease could be controlled by phytosanitation in an isolated plot away from other cocoa to avoid possible contamination. However, the large number of variables affecting disease incidence after sanitation would have made comparison between regions difficult, and such an experiment was not carried out as part of the IWBP. A second alternative was a factorial experiment to include the effects of timing and frequency of removal, sanitation ratio, shade, etc. This too was impractical owing to the area and labour required for a factorial trial.

INTERPRETATION OF RESULTS

The project generated a large database, with comparable data from all sites, covering each of the important cocoa growing regions of South America affected by the disease. The results were interpreted first in the individual countries contributing to the studies, and then decisions were made collectively by the IWBP scientists at the 6th and 7th IWBP Workshops on whether the various methods of disease control were applicable in each region. Strategies for disease management were derived from a combination of those methods that were relevant.

Chapter 4

COMPARATIVE EPIDEMIOLOGY STUDIES: INTRODUCTION

S.A. Rudgard, T. Andebrhan, A.C. Maddison & R.A. Schmidt

METHODS FOR COMPARATIVE EPIDEMIOLOGY EXPERIMENT

Establishment of Research Sites

A protocol was established for the installation of research sites and recording of data, so that results from different sites would be comparable. Initially eight sites (three in Brazil, two in Colombia, two in Ecuador, and one in Trinidad) were planned to start in the first dry season at each location after July 1985. Two more sites (one in Grenada, and one in Venezuela) were added one year later. Recording also began in 1986 at a site in Bahia, Brazil, where witches' broom was absent. Local scientists identified locations, either from within institute research stations or in commercial plantations belonging to cooperating farmers. The *site* was to consist of a small number of trees selected from within a larger mature *planting*. Ten trees within the *site* formed the *plot*, used for more detailed recording. Detailed descriptions of the *sites* were prepared and the information compiled into a standard format.

The general information on geographic location required was: latitude and longitude, elevation above sea level, topography, and location of the nearest permanent meteorological station. The following details of the *planting* were recorded: size (hectares), soil type, tree age, spacing, ground cover, types of shade tree, distance to the nearest cocoa planting, and the recent history of management of the chosen planting including any general agronomic practices. The experimental site consisted of 50 numbered trees selected from within the *planting*, from which 10 representative sample trees were chosen to form the *plot*. The area of the *planting* containing the *site* was shown on a plan, also indicating positions of shade trees and the site meteorological station. The following characteristics were recorded from each of the 10 sample trees: maximum height, trunk girth at 0.5 m, jorquette height, canopy width and depth, and genetic material where known.

Marking of Trees and Recording

Recording was divided into five artificial categories for ease of data handling as follows: 1. Vegetative shoots, 2. Flowers, 3. Fruits (pods), 4. Basidiocarps, and 5. Meteorology. Datasheets were devised and recording intervals were set so that data would be directly comparable.

Disease Management in Cocoa: Comparative epidemiology of witches' broom. Edited by S.A. Rudgard, A.C. Maddison and T. Andebrhan. Published in 1993 by Chapman & Hall, London. ISBN 0 412 58190 6

For registers of growth of vegetative shoots, five units each of at least 10 healthy growing points were selected and marked on the 10 plot trees. The units were intended to be representative of the canopy as a whole. The following characters were recorded every 2 weeks in each unit: number of actively growing "flush" shoots, total number of immature leaves, and the number of infected shoots in six categories. These last were: swellings/hypertrophies on leaf peduncles or pulvinii, swellings on stems, axillary brooms associated with one or other type of swelling, axillary brooms not associated with swellings, and terminal brooms.

Flowering was registered in four 1 m lengths of branch selected and numbered at random in the lower canopy, and one 1 m length in the middle of the trunk (50 cm if the trunk was short), of the 10 plot trees. Clonal trees had no identifiable trunk or basal orthotropic shoot, so five branch lengths were selected. Flowers formed from a floral meristem, known as a "cushion". The following characters were recorded within these lengths every 2 weeks: number of active flower cushions (with flower buds), number of active cushions with open flowers, and number of newly-infected cushions with any combinations of three symptoms: necrotic flowers (type 1), chirimoyas (type 2), and vegetative cushion brooms (type 3). Cushions were marked individually with pins when they were first observed to be flowering or infected. Infected cushions were only registered once at the first appearance of any symptoms. Pins marking active cushions were counted at the end of each year.

Fruiting was registered in two ways. Firstly, five additional 1 m branch lengths were selected and marked in each plot tree apart from the five used for flowering as above. The number of fruit in the following categories was recorded in all 10 branch lengths every 2 weeks: newly-set fruit more than 1 cm long, newly-wilted fruit, developing fruit with no necroses, newly-registered chirimoyas, healthy ripe pods, necrotic pods with infections of *Crinipellis perniciosa*, *Phytophthora* spp. or *Moniliophthora roreri*, and pods with areas of delayed ripening (green- islands). Every newly-formed fruit was marked with the date when it was first observed. Newly-set or wilted fruit, developing fruit and chirimoyas were left on trees, and fruit from the other categories were removed when observed. Secondly, fruiting was registered when pods were removed from all site trees as they became ripe or necrotic. Harvest data were recorded every 2 weeks for the 50 trees, with additional records of heights above the ground for all pods harvested from the 10 plot trees. The number of pods collected from each tree was recorded in the categories of ripe healthy, or with one or more diseases. All fruit harvested from within the marked branch lengths had the date of their first observation recorded as well. All necrotic pods in sites where *M. roreri* was present were split and left for 7 days in plastic bags to check for sporulation of that pathogen.

Basidiocarp production of *C. perniciosa* was registered as the number of open turgid basidiocarps produced on one day in each calendar week on a group of diseased objects. Records were taken early in the morning whenever possible, and if rain had fallen in the previous 18 hours. No basidiocarps were removed. Three principal types of necrotic diseased objects were evaluated as sources of basidiocarps, namely: vegetative brooms, flower cushion infections (henceforward cushion brooms), and diseased pods. Sets of detached vegetative brooms and pods were suspended from cord tied between tall poles at the top of the canopy, in the middle, and in the leaf litter on the ground for each plot tree. Diseased material was obtained at random from tree canopies within the site, and 10 objects were placed at each position in the 10 trees except for the suspended pod sets which were reduced to five. New samples of brooms and pods were placed alongside the first samples in year 2. Any suspended objects that fell were re-hung if possible or, if not, were replaced. Leaf litter was allowed to fall on to marked objects placed on the ground, and was lifted off carefully and replaced during weekly inspections. Cushion brooms were selected and marked at random through the canopies of the plot trees, but were left attached to trees. A maximum of 20 cushions per tree was identified in year 1, and another 20 in year 2. In sites with too few infected cushions, infections were detached from other trees in the site and attached to the plot trees to make up the correct total.

Meteorological data were available from established permanent stations near the plots. A set of equipment was also placed as near to the site as possible but outside the canopy of the planting, containing the following: thermo-hygrograph (Casella London), weighing bucket rain-gauge (Belfort Instruments, USA), wetness recorder with string sensor (Belfort Instruments, USA), maximum/minimum thermometer (Casella London), and a sling psychrometer (Forest Suppliers, USA). The thermo-hygrograph and thermometer were placed in a weather shelter at standard height, and the wetness recorder was placed inside its own shelter at the same height. Key data were extracted from the equipment charts for daily maximum and minimum temperatures with total duration above 30°C and below 20°C, and maximum and minimum relative humidities with duration above 95% and below 70%. Total daily rain was also found, with the start time of the first rain event and the finish time of the last. The daily maximum percentage scale deflection was found for the wetness sensor, with the time at 100% deflection, and the start and finish time of any scale deflection. The finish time was often after 24.00 h of the start day for both rain and wetness.

Further Management of Sites

Once all the infected material for registering basidiocarp production was collected, marked and put in position, all other infected brooms cushions and pods were removed from the site trees and counted before the start of the experiment. The removed material was left on the ground in the site. Infections were also removed from as much of the cocoa field as possible before the experiment. No infections were removed from any trees in the cocoa field during the experiment, and in some cases the total number of diseased points was found at the end of the experiment.

Infected and ripe healthy pods were removed from site trees as above, and pods were harvested from other trees in the cocoa field throughout the experiment as dictated by normal regional practices. Chupons were removed in the cocoa field according to local practices.

SITE PROFILES

No criteria were laid down for the selection of cocoa plantings for IWBP sites, other than that they be representative of cocoa plantations within the different regions as discussed in Chapter 3. There are several agronomic aspects important in determining the microenvironment inside a plantation which were not standardised between all locations. Each site was unique in the experiment so full descriptions were obtained before recording started. The information available falls into three main categories: location of the planting, nature of the planting, and details of the ten plot trees themselves.

TABLE 4.1
Positions of sites and rainfall characteristics

Site	Longitude (W)	Latitude	Altitude (m)	Annual rainfall (mm)[a]	No. months < 100 mm /year[1]
Altamira	53°20'	03°30' S	50	2,100	6
Tomé Açu	48°20'	02°40' S	50	2,800	4
Ouro Preto	62°10'	10°40' S	200	2,000	4
Llanos	73°40'	02°50' N	250	3,600	1
Manizales	75°40'	05°00' N	1,000	2,200	1
Pich. Station	79°30'	01°00' S	50	2,500	6
Pich. Farm	79°30'	01°00' S	50	2,500	6
Grenada	61°40'	12°10' N	50	1,900	4
Trinidad	61°20'	10°40' N	100	1,800	4
Táchira	72°00'	07°30' N	100	2,800	3

[1] Calculated from long-term normals.

Figure 4.1. Locations of sites in the comparative epidemiology experiment.

The locations of the sites are shown on the map of South America (Figure 4.1), and the latitudes and longitudes in Table 4.1. The Brazilian and Ecuadorian sites are south of the equator, and all sites except Manizales are below 300 m above sea level. Average annual rainfall varies between 1,800 and 3,600 mm at Trinidad and Llanos respectively. All sites except Manizales and Llanos have an average of at least 3 months with less than 100 mm of rain.

The sites established in Brazil were at Altamira, Tomé Açu and Ouro Preto, all less than 200 m above sea level. The first two are in the state of Pará in the lower Amazon Basin, 500 km and 200 km from the sea respectively. The terrain is flat to slightly undulating. Ouro Preto is on the southern edge of the Amazon Basin about 1,000 km south of Manaus in Rondonia state, where the terrain consists of low hills. One of the Colombian sites was located near Manizales in the montane valley of the river Chinchina, and was the highest of all at approximately 1,000 m above sea level. The other was installed on the Llanos plains east of the Andes and close to Villavicencio, in flat terrain close to a river (Ari–Ari). The Venezuelan site was located in Táchira State on the plains south–east of the Andes, in a very similar location to the Colombian Llanos by a river. The last two sites were approximately 300 m above sea level and 500 km apart. The two Ecuadorian sites were separated by only 5 km, situated close to Pichilingue Quevedo on the coastal plain about 50 m above sea level. The remaining two sites were on Caribbean islands, namely Trinidad and Grenada, at most 10 km from the sea.

TABLE 4.2
Type and number of shade trees per hectare in sites

Site name	Principal genera of shade trees	Number of live shade trees	Shade trees /ha
Altamira	none	0	0
Tomé Açu	*Erythrina*	45	110
Ouro Preto	*Sweitenia*	12	50
Llanos	*Inga, Gliricidia*	21	500
Manizales	*Erythrina, Samanea, Inga*	6	40
Pich. Station	*Inga*	6	30
Pich. Farm	mixed	8	60
Grenada	*Gliricidia*	1	10
Trinidad	*Erythrina*	2	30
Táchira	*Erythrina*	17	150

All sites except Altamira were in plantings with some degree of overhead shade provided by one or more of the common species. The shade trees had been planted at regular intervals in some sites, but death of some often resulted in irregular patterns. Canopy spread depends on species and age but a rough index of planting density was calculated using the plot area and numbers of shade trees of any type (Table 4.2). It can be seen that the Llanos, Táchira, and Tomé Açu sites all had high densities. Ouro Preto and both Pichilingue sites had moderate densities, whilst in Trinidad and Grenada sites the shade was sparse and patchy because only one or two trees were present.

Spacing between the cocoa trees in the plots varied considerably between 2 m square and 4 m square (Table 4.3). Spacing between the cocoa trees had a large influence in most sites over height and width of individual trees. Exceptions were the two Pichilingue sites where cocoa was exceptionally tall and the Llanos site which was very short, the latter possibly due to immaturity of the trees and poor soil. However, the more widely spaced plantations generally contained shorter trees with wider canopies, and the older trees had greater trunk diameters. Grenada, at 30 years, was the oldest planting and the Llanos was the youngest at six years.

TABLE 4.3
Details of cocoa trees in sites and mean measurements of plot trees

Site name	Site				Plot[1]		
	Planting material[2]	Tree age (yr)	Spacing distance (m)	% dead trees	Tree height (m)	Canopy width (m)	Trunk diameter (cm)
Altamira	S many	9	3 x 3	17	5.1	4.6	13.6
Tomé Açu	S many	7	2.5 x 2.5	19	5.6	5.4	14.8
Ouro Preto	S two	13	4 x 4	16	4.5	5.1	17.8
Llanos	S many	6	2 x 2	3	3.1	3.9	7.9
Manizales	C one	18	4 x 4	32	4.8	5.6	20.7
Pich. Station	S one	19	4 x 4	5	6.4	5.4	18.7
Pich. Farm	S many	16	3.5 x 3.5	7	7.5	4.9	15.3
Grenada	C many	30	3 x 3	19	5.6	4.2	19.7
Trinidad	C two	10	2 x 4	15	4.1	3.6	12.3
Táchira	S many	25	4 x 4	20	4.4	5.7	14.8

[1] 10 trees per plot; [2] S, seedlings; C, clones; many/two/one, number of clones or crosses.

Genetic material in the sites was not standardised, particularly as heterogeneity was great within plots on commercial farms. All materials were considered by the local scientists to be moderately to highly susceptible to witches' broom disease. Trees grown from rooted cuttings of identified clones made up the sites at Manizales (SC 6), Trinidad (TSH 1188 & 919) and Grenada (ICS clones). The site at Ouro Preto (Catongo x UF613 & Sca 12) and the Station at Pichilingue (EET95 x Silecia 1) were of one or two identified hybrids grown from seed. The Farm site at Pichilingue comprised a mixed planting of unidentified Trinitario/Nacional open-pollinated crosses. The plots at Altamira, Tomé Açu, Llanos, and Táchira were all in plantations, mostly commercial properties, of between ten and twenty hybrids planted as seed mixtures where individual trees could not be identified with certainty. Most trees in the two Brazilian sites were progeny of Catongo, IMC67 and UF clones.

The protocol for the experiment required that all brooms be removed as stated above, and counted, though brooms were not removed at Pichilingue Farm at all in order to recreate normal farmer practices in that locality. Material removed from trees was classed either as cushion or vegetative brooms. Ouro Preto and the two Pichilingue sites had the largest numbers of cushion brooms per tree, but Ouro Preto also had large numbers of vegetative brooms (Table 4.4). The Pichilingue and Manizales sites were similar in that they had many more cushion than vegetative brooms. Altamira, Tomé Açu, Manizales and Táchira were moderately infected, and Grenada, Trinidad and the Llanos had few brooms per tree.

TABLE 4.4
Mean number of brooms/tree removed from plot trees at IWBP sites

Site name	No. brooms removed	
	Vegetative	Cushion
Altamira	57	46
Tomé Açu	166	25
Ouro Preto	234	262
Llanos	<5[1]	<5[1]
Manizales	10	95
Pich. Station	43	421
Pich. Farm	18[2]	179[2]
Trinidad	19	14
Táchira	ND	ND
Grenada	<5[1]	<5[1]

[1] visual estimates; [2] counts at end of trial; ND no data available.

Chapter 5

COMPARATIVE EPIDEMIOLOGY EXPERIMENT: BRAZIL

T. Andebrhan, J.C.B. Costa, A.L.P. Carvalho & P. Albuquerque

INTRODUCTION

Importance of Cocoa and Witches' Broom in the Brazilian Amazon

Cocoa has been cultivated in the Brazilian Amazon region for about 250 years. The original plantings were concentrated mainly on the banks of the River Amazon between Manaus and Belém, much of it in the low–lying areas that are flooded for a large part of the year.

The Federal Government decided in the early 1970s to encourage settlers to plant cocoa in the Amazon region by setting up a rural credit scheme, consisting of a package of financial and extension/research support provided by CEPLAC (Comissão Executiva do Plano da Lavoura Cacaueira), the Brazilian government agency responsible for cocoa. A plan to extend cocoa cultivation in Brazil called PROCACAU was started in 1975. The new areas of colonisation throughout the Amazon region all had efficient credit and extension support in operation for almost 10 years, and so by 1985 the planted area had reached almost 100,000 ha (Table 5.1). The total crop production in Amazonia until 1979 was about 1,500 t (mainly from the Lower Amazon region). From then, it increased rapidly as the new plantings came into bearing, so that by 1990 it had risen to about 80,000 t. The Amazon region is characterised by comparatively high yields for South America of about 750 kg/ha, owing to the priority being given to planting in areas of fertile soils and use of planting material with higher potential yields.

The first outbreaks of witches' broom in the cultivated crop would have been initiated by inoculum from the indigenous wild cocoa population in the newly colonised areas such as the State of Rondônia. However, the disease had probably always been present at very low levels in traditional cocoa areas of the lower Amazon. Witches' broom was recognised as a serious problem in Rondônia in about 1978, and it was widespread by 1982 due to lack of disease management, provoked in turn by the low international cocoa price. The land area under cocoa in that state was sufficiently large by 1980 that the rate of spread to uninfected plantations had substantially increased.

An evaluation of witches' broom in the four main planting regions of Rondônia by CEPLAC in 1982 showed that the disease was most serious in the oldest cocoa. A disease

Disease Management in Cocoa: Comparative epidemiology of witches' broom. Edited by S.A. Rudgard, A.C. Maddison and T. Andebrhan. Published in 1993 by Chapman & Hall, London. ISBN 0 412 58190 6

control campaign was started in 1983 which consisted of an education programme to increase growers' awareness of the disease and its control and a massive effort to train farmers and labourers in broom removal (phytosanitation). The campaign coincided with a large rise in the price of cocoa, and the response from growers was substantial. The campaign continued until 1985, by which time the majority of farms regularly used phytosanitation to reduce pod losses. Economic conditions were still favourable, and cocoa was a highly profitable crop. The area of cocoa continued to expand up to 1986, when the rural credit schemes became less available and less cost-effective.

TABLE 5.1
Estimated areas (ha) planted with cocoa in the Brazilian Amazon Region

State	1975	1980	1984
Acre		400	2,000
Amazonas		800	1,000
Maranhão		300	500
Mato Grosso		2,500	4,000
Pará	1,000	17,500	45,000
Rondônia	400	23,500	50,000
Estimated total	1,400	45,000	102,500

Source : Alvares-Afonso, 1984

Cocoa prices began to fall at the end of 1985, and the general response of growers was to manage the crop less intensively and relax sanitation practices. Witches' broom incidence began to increase, and became severe throughout most of the Amazon region. New plantings were few, and concentrated in areas of colonisation of primary forest such as southern Pará State.

Seed material of varieties planted in Amazonia was derived from clonal seed gardens, initially from CEPLAC, Bahia, and latterly from CEPLAC stations in Amazonia. Clonal parents were chosen from the standard commercial varieties from Trinidad (ICS accessions), the Pound collection (IMC, PA, P & Sca accessions) and Costa Rica (UF). The reactions to witches' broom of most of the above were already known from other countries, and there was a proportion of crosses with the resistant Scavina (Sca) clones as one parent in the mixed seed lots distributed to farmers. The only native Brazilian cocoas used were the Catongo and "Comum" varieties. The early epidemic was accelerated by the highly susceptible progeny of parents such as Catongo and UF clones, and these were later dropped from the seed production programme.

Witches' Broom Research in Brazil

A specific team entirely dedicated to studying the agronomy of cocoa in the Amazon region was set up early in PROCACAU in Belém, Pará State. Five regional research stations were then established in addition to the headquarters and germ-plasm collection in Belém itself. Research covered all aspects and constraints of cocoa cultivation, and witches' broom was one of the most important priorities. Plant pathologists developed some knowledge of the disease and its management in the Amazon region, and kept close contact with the associated extension services. However, IWBP was seen as a means of providing comparable data for sites in the three main producing areas of the region that were already known to have different disease potentials, thereby providing the means to make management practices more effective for each region.

Current Practices for Disease Management of Witches' Broom

The only recommended practice to control witches' broom at that time was the removal by pruning of all infected parts at specific times of the year. Sanitation was normally combined with the annual structural pruning to manage the canopy, and most growers used the recommended practices provided suitable economic conditions existed. General agronomic practices such as weeding and chupon removal were maintained continuously whilst fields were in production. Some growers applied insecticide (endosulfan) once annually to control mirids and thrips when attacks were bad or prices favourable or both. The current economic situation of the cocoa industry means that phytosanitation has virtually ceased on cocoa farms in the Brazilian Amazon.

METHODS

Description of the Brazilian Sites

Altamira

The areas of cocoa cultivation along the Transamazonica highway around Altamira represented the main cocoa-producing region of Pará State. The crop was introduced in the early 1970s, and in contrast to the rest of the Amazon region, the development of witches' broom had been very slow. Canopy infection was often comparable to other localities, but pod losses were lower.

The experimental plot was located in a field of 0.75 ha of unshaded cocoa on a commercial farm, about 5 km from the CEPLAC station and 300 m from the nearest cocoa plantation, where witches' brooms had been removed once in 1982. The IWBP site (0.25 ha) was cleared of brooms in 1985, and the remaining 0.5 ha was not pruned. Data recording started in late November (week 48) 1985, and continued to December (week 52) 1987. The IWBP meteorology station was not installed at the site until February 1986, and meteorological data prior to this period were taken from the CEPLAC station.

Ouro Preto

The site was in Ouro Preto in Rondônia State, where pod losses during the peak harvest (May–July) were known to reach 70–80% in some farms even after application of phytosanitation practices. The experimental site was part of a 4 ha field at the CEPLAC experimental station, planted as a spacing trial with 4 hybrids in 1976. Brooms had been removed annually until 1982, and the trees were severely infected by the start of IWBP in 1985. The trees' canopies were affected by the first broom removal, so the experiment did not start until 1986 to allow the trees to recover. All neighbouring cocoa was cleared of brooms. Data recording ran from September (week 34) of 1986 to September (week 32) 1988.

Tomé Açu

The site was located near the town of Tomé Açu in southern Pará State, where cocoa was cultivated principally by the Japanese immigrants who settled there around 1930. They introduced black pepper to the Pará State, which became the largest producer in Brazil. An epidemic of root disease of black pepper caused by *Fusarium solani* in the late 1970s forced the farmers to diversify their crops. The soil in Tomé Açu was of low natural fertility, and cocoa was only chosen as an alternative crop to take advantage of the residual fertilizers. It was treated like a horticultural crop returning high yields (2,000 kg/ha) with high inputs, and witches' broom was managed through effective sanitation practices keeping pod disease to about 10–20%.

The experimental site at Tomé Açu was located in a 1 ha cocoa planting of mixed hybrids established in 1979 at the old Japanese government's experimental station, INATAM (Institute of Amazonian Tropical Agriculture), which had been handed back to the Brazilian government in 1985. It was 300–400 m away from the nearest cocoa farm, and had not been sanitised before IWBP. Data recording started in September (week 38) 1985, and continued to December (week 53) 1987.

Site Installation

In all the sites, the 10 trees for IWBP studies were selected for their apparent susceptibility to *Crinipellis perniciosa*, and average height and canopy size. The required number of infected flower cushions (20/tree) was marked, and all diseased tissues were removed and counted. The sets of brooms and pods were placed in position in the canopy and on the ground, and all the other diseased tissues were left on the ground. No sets of already fallen brooms were used as detailed in the protocol. Diseased pods for basidiocarp counts in Altamira and Tomé Açu were collected from other sites. Sets of vegetative brooms for basidiocarp counts were taken from existing necrotic brooms already available in the sites, except in Ouro Preto where only green brooms were used. A second removal was completed two months later in accordance with the recommended practices in the Brazilian Amazon.

Units were set up with 10 healthy vegetative growing points per unit. All the points within the units were numbered, and flushing was recorded separately for each terminal point and the new growth that it subtended. All flower cushions were also numbered, and activity was registered separately for each one. General aspects of harvesting and maintenance of plantations were continued at all sites. No brooms were removed in the experimental sites at the start of the second year of the study. All other methods for recording data followed those laid out in Chapter 4.

RESULTS AND DISCUSSION

General Climate

The rainfall pattern in the Brazilian Amazon is normally well-defined. Long-term averages from Tomé Açu and Altamira show that a rainy season occurs in the period between

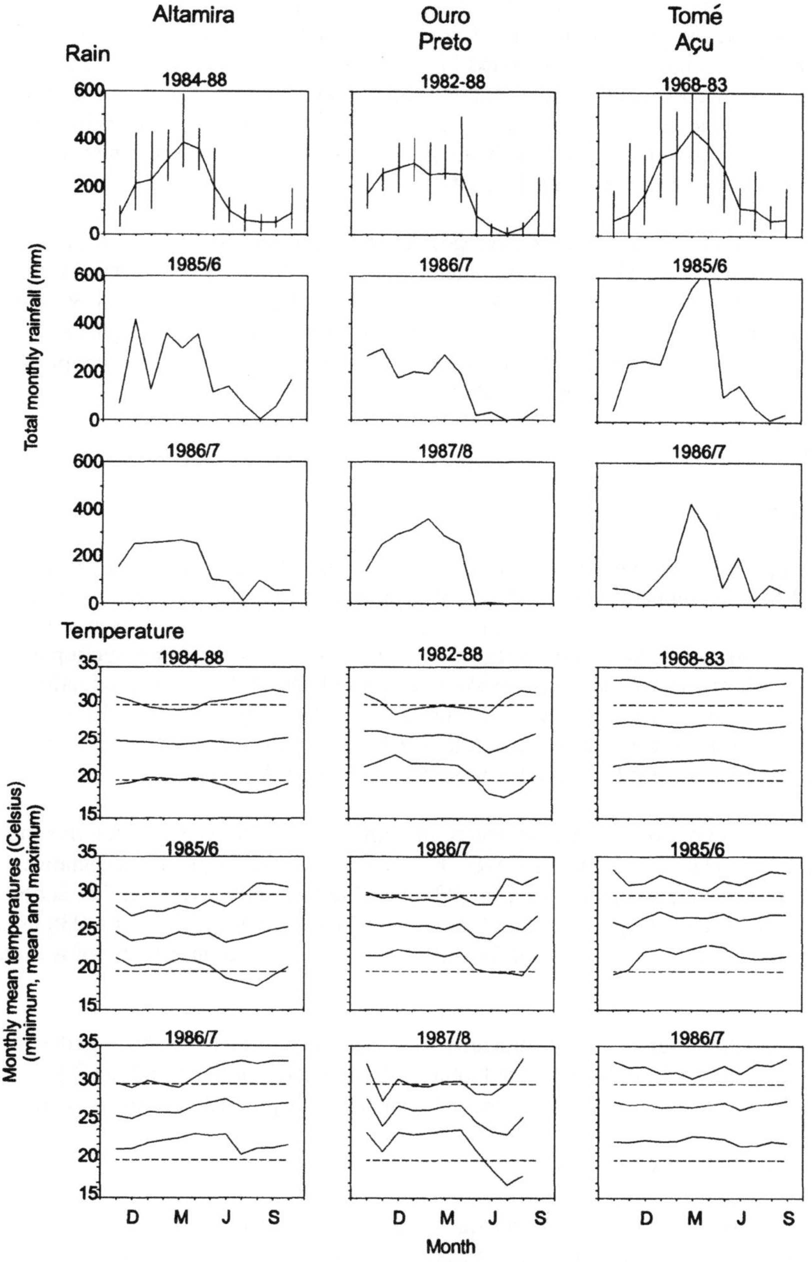

Figure 5.1. Rainfall and temperature data compared with long-term averages from the Altamira, Ouro Preto and Tomé Açu Sites, Brazil.

December and May, and from Ouro Preto that the rainy season is between October and April. In the first year, the greatest quantity of rain fell in Tomé Açu (2,768 mm), and the least in Ouro Preto (1,715 mm). The dry period in all the sites lasted 5–6 months (Figure 5.1). The IWBP data showed broadly typical patterns of wet and dry periods in the 2 years of the experiment in relation to the long-term data. The first year was much wetter in Altamira and Ouro Preto, with November and March being the driest and wettest months respectively. Less rain fell in the second year in all the sites.

Long-term air temperature averages showed little seasonal variation in all the sites, with Tomé Açu having the highest mean temperature and Altamira the lowest. Temperatures below 20°C were often recorded in Altamira and Ouro Preto during the dry season (Figure 5.1). The three variables, maximum, mean and minimum, in the 2 years of the experiment in Tomé Açu and Ouro Preto were mostly comparable with the long-term temperature data. Means in Altamira were higher than normal.

Altamira

Basidiocarp production

Basidiocarps were first observed in week 5 of 1986, after 9 weeks with rain (Figure 5.2). A total of 1,589 basidiocarps was produced on suspended vegetative canopy brooms during the 1985/6 rainy season, and the basidiocarp index[1] was 8.0. There were 21 active weeks[2], and production was concentrated into weeks 5 – 21 of 1986. Basidiocarp production restarted in the first week of the second rainy season (week 48 of 1986), but most production occurred later in the season between weeks 1 and 19 of 1987 (Figure 5.3). Total basidiocarp production fell in the second wet season to 721 basidiocarps with a basidiocarp index of 3.6, although brooms were productive for more weeks in year 2.

The second most productive source in both experimental years, vegetative brooms on the ground, produced basidiocarps over the same period as suspended vegetative brooms. There were 17 active weeks on Set 1 in 1985/6, and only 1 active week on that Set in 1986/7. Set 1 ground brooms only produced 3 basidiocarps in the second year against 238 in the first. Set 2 ground brooms were active for 24 weeks in 1986/7, and produced 19% of the total from all sources.

The other sources of basidiocarps (diseased cushions and pods) together produced between 8.5, and 15% of the combined totals of basidiocarps of Set 1 and Set 2 in both years. Diseased pods on the ground in the second year failed to produce any basidiocarps.

Flushing and vegetative infection

In year 1, there were 5 shoot growth flushes at intervals of 8–10 weeks, except for the 16 week interval after the second flush. Peaks of flush activity in year 2 were similar to year 1 with a forward shift of 2 weeks. Two additional minor peaks were recorded (Figure 5.3).

---

[1] basidiocarp index : mean number of basidiocarps per object

[2] active week : week in which basidiocarps were observed on brooms

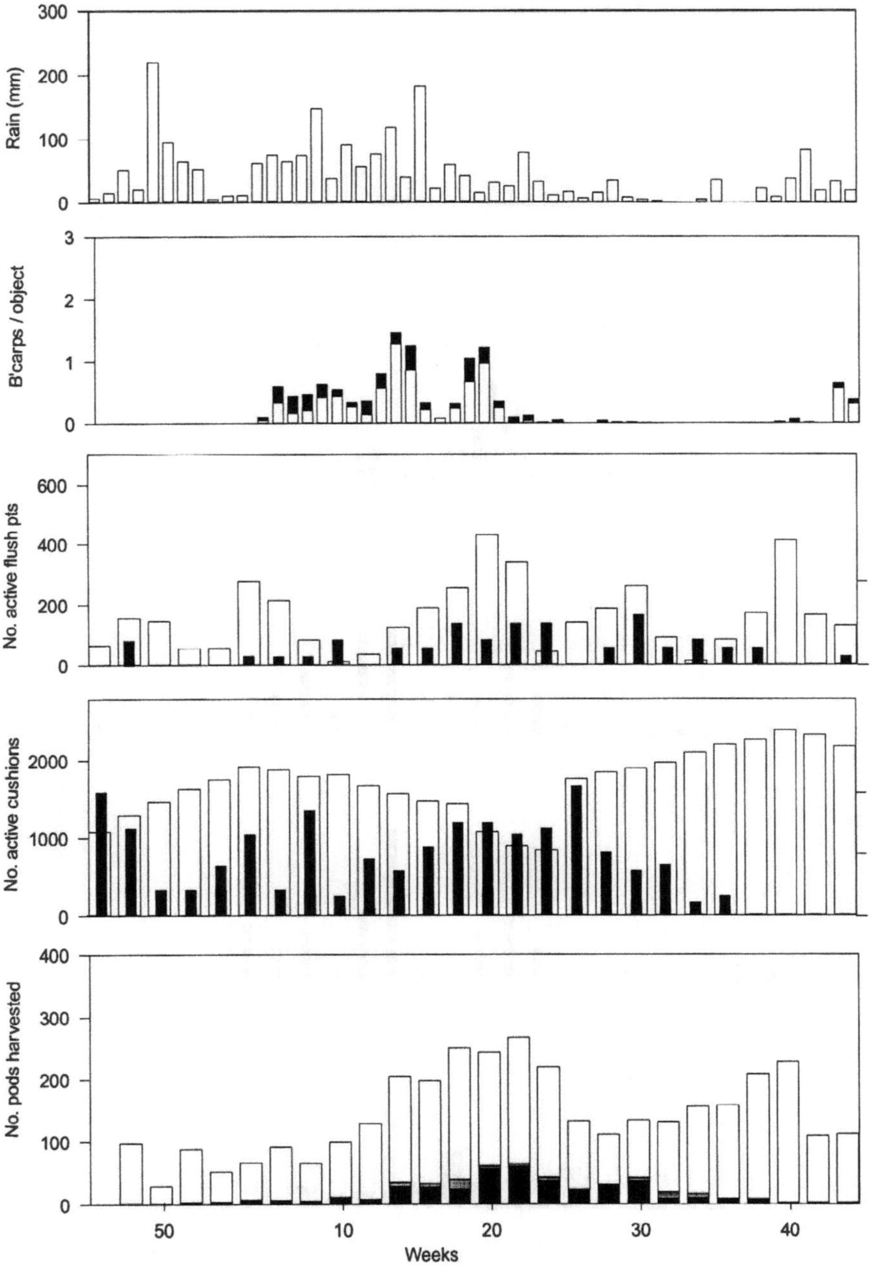

Figure 5.2. Rainfall, basidiocarp production[1], vegetative flushing and infection[2], flower cushion activity and infection[3], fruiting and fruit infection[4] in the Altamira Site, Brazil during 1985/6.

[1] open columns, the most productive source type and closed columns, sum of the rest; [2] open columns, no. of healthy flushes and closed columns, no. of infections; [3] open columns, no. of active healthy cushions and closed columns, no. of infections; [4] closed columns, witches' broom infected, shaded columns, other diseases, and open columns, mature healthy fruit.

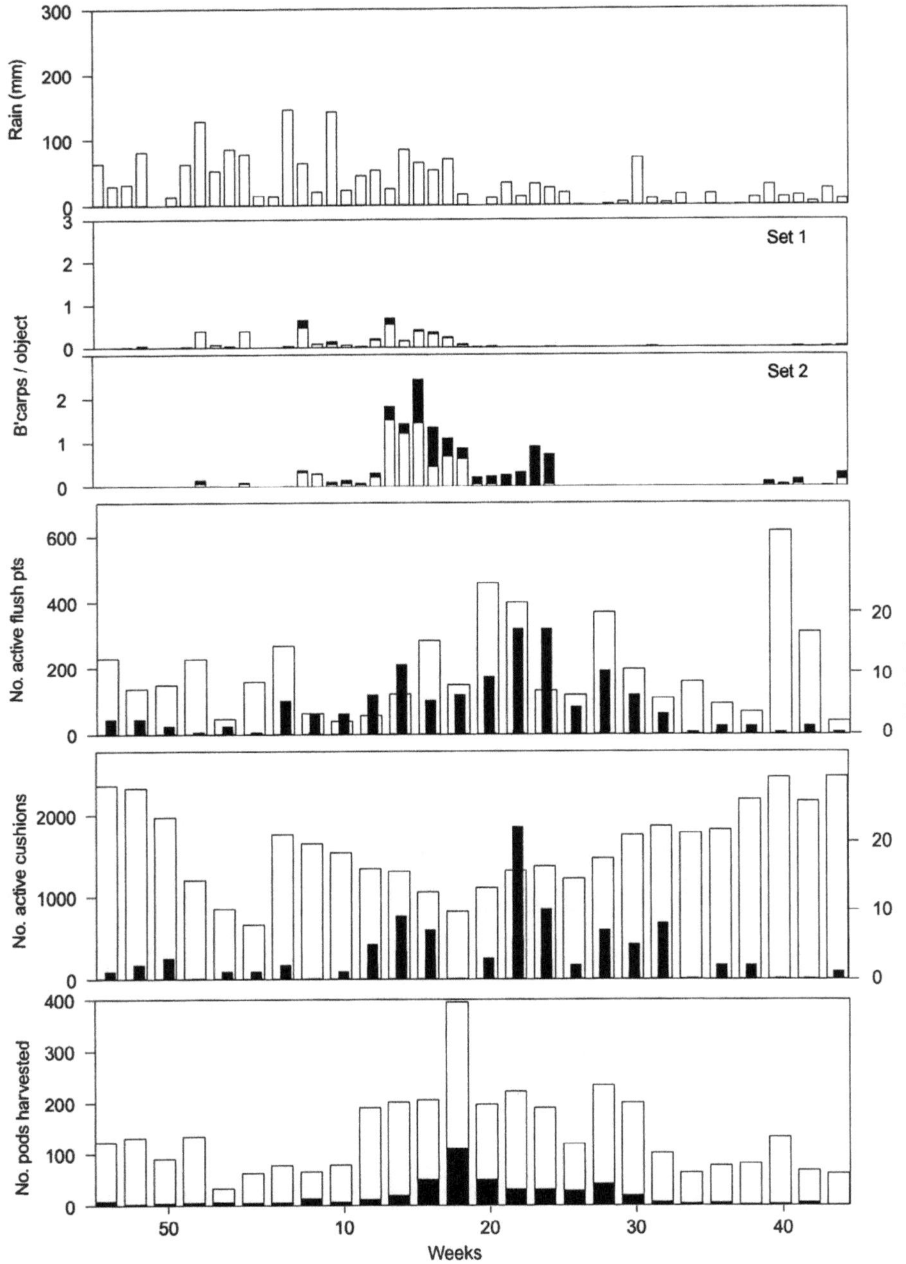

Figure 5.3. Rainfall, basidiocarp production[1], vegetative flushing and infection[2], flower cushion activity and infection[3], fruiting and fruit infection[4] in the Altamira Site, Brazil during 1986/7.

[1] open columns, the most productive source type and closed columns, sum of the rest; [2] open columns, no. of healthy flushes and closed columns, no. of infections; [3] open columns, no. of active healthy cushions and closed columns, no. of infections; [4] closed columns, witches' broom infected, shaded columns, other diseases, and open columns, mature healthy fruit.

All the weeks of peak flush activity were also associated with peaks of production of new growing buds. The increase in flush activity may have been related to the lower rainfall in year 2. It was noted in Altamira that mirid (*Monalonion* sp.) attack was severe during the rainy season. This caused death of young shoots and stimulated proliferation of new buds below the dead shoots. The increase in the number of growing points might have exposed a greater quantity of tissue to inoculum.

A significant amount of shoot activity in the two years coincided with the inoculum period from the broom sets. Growth flushes occurred at the beginning and end of the inoculum period in year 1, and infections arising during the two main flushes in weeks 6 to 20 contributed greatly to the season's total of disease symptoms. About 45% of all disease symptoms were recorded in weeks 16–26, 1986, which coincided with part of the main period of basidiocarp production (weeks 8–20), allowing for an incubation period of 4–6 weeks. The recorded inoculum period in year 2 was longer, and included most of the major shoot flushes. The overall incidence of vegetative disease increased greatly (133%) in the second year. The greatest incidence of vegetative disease occurred 4–8 weeks after peaks of flush activity, and about 30% of the symptoms occurred in weeks 24 and 26, 1987.

Flowering and cushion infection

Cushion activity in year 1 was intense until weeks 22–26 of 1986 when it fell (Figure 5.2). There was a similarly timed decrease in flowering in year 2, together with another built in activity around week 5 (Figure 5.3). A mean of 13 new cushion meristems became active on each tree in year 2.

Basidiocarps were not seen on the sample sets of brooms until week 6 of 1986, so the cushion disease recorded up to week 12 could hardly be accounted for by observed basidiocarp production. The early symptoms must either have been latent infections from the previous rainy season in 1985, or be related to sources of inoculum in the site that started to produce basidiocarps before the sample sets. Most cushion disease in year 2 could have been derived from infection in the observed inoculum period. There was no statistical relationship between the incidence of basidiocarps and the incidence of diseased cushions allowing for a range of incubation times by lagging basidiocarp data. Some periods of greater incidence could be related visually to basidiocarp production from the graphed data, for example the disease outbreak in weeks 22–24 of year 2 (1987) with the main peak of basidiocarp production in weeks 14–17.

Flower brooms comprised 76.6% of all diseased cushions, and vegetative cushion brooms observed singly or associated with other classes of disease symptoms comprised 14.9%. Flower brooms (54%) were the principal disease symptoms in year 2, followed by cushions with vegetative brooms (26.6%).

Fruiting and fruit infection

The majority of the pods completed the first 12–14 weeks of their development, when they were susceptible to infection by *C. perniciosa* (Andebrhan, 1981), during the weeks when basidiocarps were observed. Most records of pods diseased with witches' broom were confined to the period from week 16 to 36, 1986 (Figure 5.2). Pods which were formed after week 24 developed in the almost complete absence of basidiocarps on the sets. Peak incidence of pod disease in weeks 22–30 was related to infections on pods exposed to the basidiocarp production in weeks in 14–20. Witches' broom disease infected 11% of pods both from within the branch units and on the larger sample of 50 trees.

Pod set in year 2 was at similar times to that in year 1, although the number of pods set in sample units declined by 21%. Pod disease in the 50 site trees was almost the same. Peak incidence of diseased pods was in weeks 14–30 in year 2, whereas the main period of basidiocarp production on sample broom sets was from weeks 5–21. Pods exposed during this period represented 52% of the total year's pod set with 16.2% becoming diseased with witches' broom. The rest of the pods were formed during the drier weeks, and completed the first 12–14 weeks of their development when basidiocarps were either absent or scarce.

Ouro Preto

Basidiocarp production

Diseased object sets used to assess basidiocarp production were formed during the dry season of 1986 (see Methods section.), and vegetative brooms were collected before they became necrotic. A total of 24 weeks with more than 5 mm of rain was required before any basidiocarps formed on suspended brooms, which occurred in week 6 of 1987 (Figure 5.4). There were only 9 active weeks in that year before basidiocarp production stopped in week 18. The same brooms were exposed to only 6 weeks of rain after the dry season in 1987 before the first basidiocarps were produced. Set 2 suspended vegetative brooms were exposed to 16 weeks of rain before basidiocarps started to form. Production started in week 2 of 1988 and continued to week 17 (Figure 5.5). The basidiocarp index was 2.8.

The second most productive source in both years, the diseased cushions, produced basidiocarps at similar times to the suspended vegetative brooms, though more formed later in the active period. Cushions were active for 6 weeks in year 1, with a basidiocarp index of 0.36. In year 2, both sets of cushion brooms were active for 9 weeks with indices of 0.55 and 0.45 for Set 1 and 2 respectively.

Diseased pods in all positions and brooms on the ground from Set 1 failed to produce any basidiocarps in either year. These same objects in Set 2 did produce some basidiocarps, but were active for less than 4 weeks.

Flushing and vegetative infection

There were 6 shoot flushes in year 1 at intervals of between 4 and 14 weeks. The intense flush 4 weeks after the start of the experiment was attributed to the initial pruning (Figure 5.4). Other flushes were recorded in weeks 44–48 of 1986, and weeks 4, 16 and 30 of 1987. Only 4 shoot flushes occurred in year 2 (Figure 5.5), with intervals of 6–22 weeks. The flushes in week 40, 1987 and week 1, 1988 represented 40% of the total number of active points recorded in year 2.

Production of basidiocarps on the sets in year 1 was concentrated into weeks 6–18 of 1986, and only one main shoot flush in week 16 was exposed to infection. Most diseased

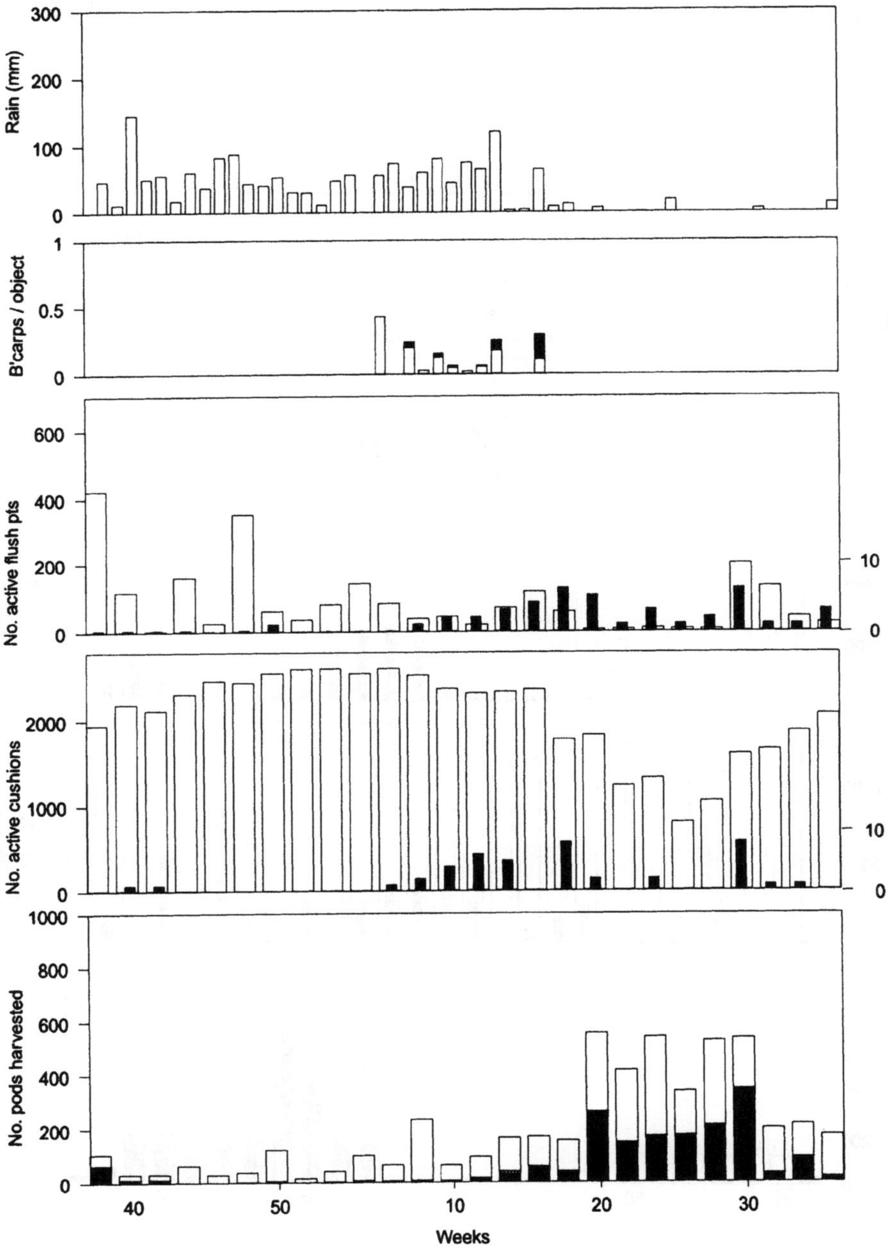

Figure 5.4. Rainfall, basidiocarp production[1], vegetative flushing and infection[2], flower cushion activity and infection[3], fruiting and fruit infection[4] in the Ouro Preto Site, Brazil during 1985/6.

[1] open columns, the most productive source type and closed columns, sum of the rest; [2] open columns, no. of healthy flushes and closed columns, no. of infections; [3] open columns, no. of active healthy cushions and closed columns, no. of infections; [4] closed columns, witches' broom infected, shaded columns, other diseases, and open columns, mature healthy fruit.

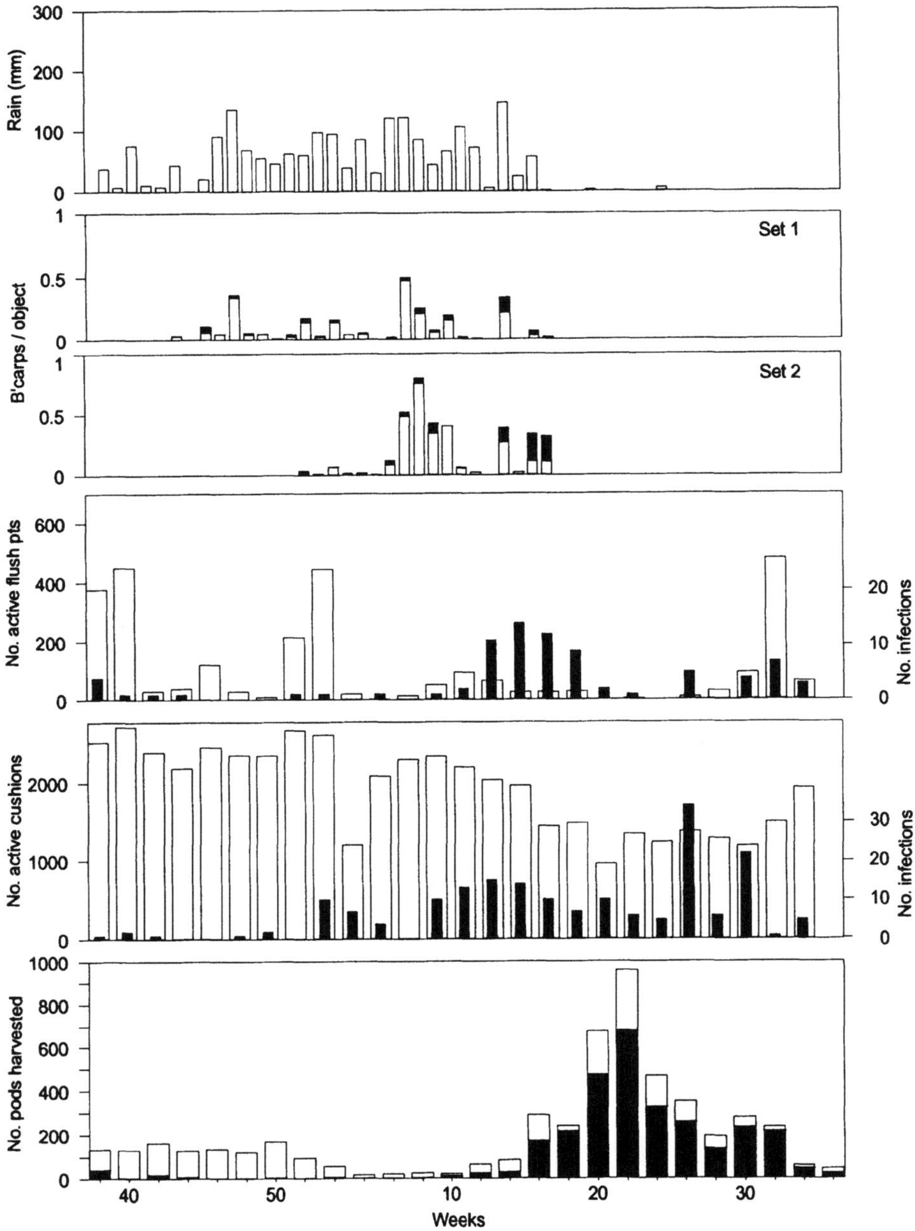

Figure 5.5. Rainfall, basidiocarp production[1], vegetative flushing and infection[2], flower cushion activity and infection[3], fruiting and fruit infection[4] in the Ouro Preto Site, Brazil during 1986/7.

[1] open columns, the most productive source type and closed columns, sum of the rest; [2] open columns, no. of healthy flushes and closed columns, no. of infections; [3] open columns, no. of active healthy cushions and closed columns, no. of infections; [4] closed columns, witches' broom infected, shaded columns, other diseases, and open columns, mature healthy fruit.

shoots appeared after week 8, and 61.5% were concentrated into the weeks 8–22, 1987, corresponding to infections arising in the main period of basidiocarp production. The flushes in year 2 (1987/8) all occurred during the period of basidiocarp production (weeks 44 of 1987 to 18 of 1988). About 61% of diseased shoots formed between weeks 13 and 19 of 1988. Diseased shoots recorded during the early rains for week 50 of 1986 and weeks 38–44 of 1987 were probably derived from infections carried over from the previous rainy season.

Flowering and cushion infection

Cushions remained highly active throughout most of year 1 (1986/7), with reduced activity between weeks 18 and 26 in 1987 (Figure 5.4). Flowering was most intense in the last quarter of 1987, around the start of the rainy period in year 2 (Figure 5.5). A total of 34 new cushion meristems was recorded in the sample branch units in year 2.

The extended period of cushion activity included the periods of basidiocarp production. Most disease symptoms appeared between weeks 12 and 30 of year 1 (1987). The weeks of greatest incidence of new symptoms in year 2 could be traced back to peaks of basidiocarp production. Infections might have remained latent from the end of the inoculum period in year 1.

Flower brooms predominated (82%) and 9% of diseased cushions with vegetative brooms were associated with other classes of symptoms, mainly with cushions with flower brooms and chirimoyas. Flower brooms and diseased cushions with vegetative brooms represented 73 and 19% respectively of diseased cushions recorded in year 2.

Fruiting and fruit infection

Some fruit set occurred throughout year 1 in Ouro Preto, but most was concentrated into the period from week 46 of 1986 to week 6 of 1987 (Figure 5.4). The first 12–14 weeks of development of these pods occurred during the period of basidiocarp production, and they accounted for 52.8% of diseased pods recorded in year 1. The mean annual pod loss to witches' broom from the 50 trees was 36.9%.

Many pods (49%) in year 2 matured in the period from week 17 to 25 of 1988. These harvests were composed of pods set between week 48 of 1987 and week 3 of 1988 which remained susceptible during the whole period of basidiocarp production in year 2. Most witches' broom diseased pods arose in the period from week 19–23 corresponding to the main inoculum period, and 53.4% of all pods became diseased during the year. Incidence of *Phytophthora* pod rot was low in year 1, and was zero in year 2.

Tomé Açu

Basidiocarp production

The first basidiocarps were formed in week 4 of 1986 on vegetative brooms at the top of the canopy (Figure 5.6) after 15 weeks with rain. Suspended brooms produced 1,422 basidiocarps, about 87% of the basidiocarps produced by all sources, during the 1985/6 season. Suspended vegetative brooms continued to produce basidiocarps until week 23 of

23 of 1986, and basidiocarp production restarted in week 48 of 1986, after a lapse of 24 weeks including 7 weeks without any rain (Figure 5.7).

Only suspended vegetative brooms produced basidiocarps in Set 2. Basidiocarp production started in week 9 of 1987 after an induction period of 27 weeks which included 11 weeks with rain, and brooms were only active for 8 weeks in all. Only 91 basidiocarps were produced, which was 8% of the total produced by Set 1. The difference could not be attributed to the quantity and distribution of rain, as those factors affected both sets of brooms equally.

The second most productive source of basidiocarps in year 1, the diseased flower cushions, produced 5.5% of the combined total of basidiocarps with a basidiocarp index of 0.45. Basidiocarps were principally observed on cushions with vegetative brooms and chirimoyas, and only rarely on those with flower brooms. Basidiocarp production was negligible in the second year on both sets of cushion brooms. Suspended diseased pods started to produce basidiocarps in week 8 and continued until week 23. Basidiocarp indices on pods were 0.84 and 0.58 in the first and second year, respectively.

Set 1 vegetative brooms placed on the ground produced 39 basidiocarps in year 1, of which 25 formed in one week (week 23) on only 4 brooms. None formed in year 2 on either set of ground brooms, and all diseased pods placed on the ground failed to produce any basidiocarps at all.

Flushing and vegetative infection

An intense period of shoot growth in weeks 42 and 44 occurred just after the pruning at the start of the experiment, during weeks in which no basidiocarps were observed (Figure 5.6). There were 4 other flushes of shoot growth in year 1 which occurred at an interval of 8–10 weeks (weeks 06, 16, 22 and 30 in 1986). After 7 weeks without rain in the dry season of 1986, a small flush of shoot growth was recorded in week 38, 1986. Other larger but more dispersed flushes were in weeks 46–50 and weeks 22–28 of 1987 (Figure 5.7). The early part of the latter flush may have been exposed to inoculum, according to the sample broom sets.

Diseased vegetative shoots appeared in two distinct groups in year 1. The first group (weeks 42–50, 1985) was not associated with previous basidiocarp production during the experiment, but was probably derived from latent infections in buds carried over from the previous rainy season. These were manifested during the strong flushing after the pruning. The other group of symptoms formed after week 20 of 1986, and arose directly from infections during the inoculum period in year 1. The minimum incubation period was about 6 weeks. Diseased shoots in year 2 were also mostly recorded after week 20.

Flowering and cushion infection

The number of active cushions was generally greater from about week 40 up to and between weeks 10 and 20, followed by a period of general inactivity in the late rainy and early dry seasons (Figures 5.6 and 5.7). There were deviations from this general trend, and the general level of activity was less in year 2, in the second and third quarters.

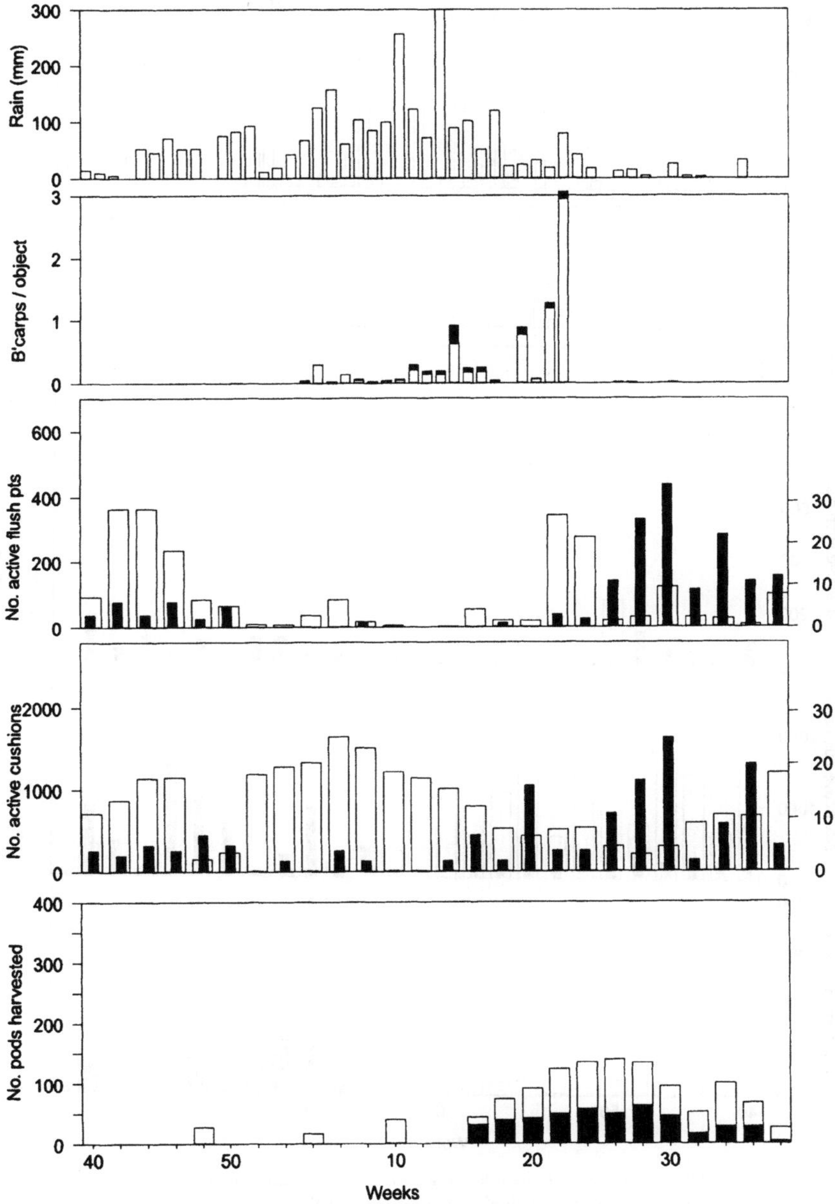

Figure 5.6. Rainfall, basidiocarp production[1], vegetative flushing and infection[2], flower cushion activity and infection[3], fruiting and fruit infection[4] in the Tomé Açu Site, Brazil during 1985/6.

[1] open columns, the most productive source type and closed columns, sum of the rest; [2] open columns, no. of healthy flushes and closed columns, no. of infections; [3] open columns, no. of active healthy cushions and closed columns, no. of infections; [4] closed columns, witches' broom infected, shaded columns, other diseases, and open columns, mature healthy fruit.

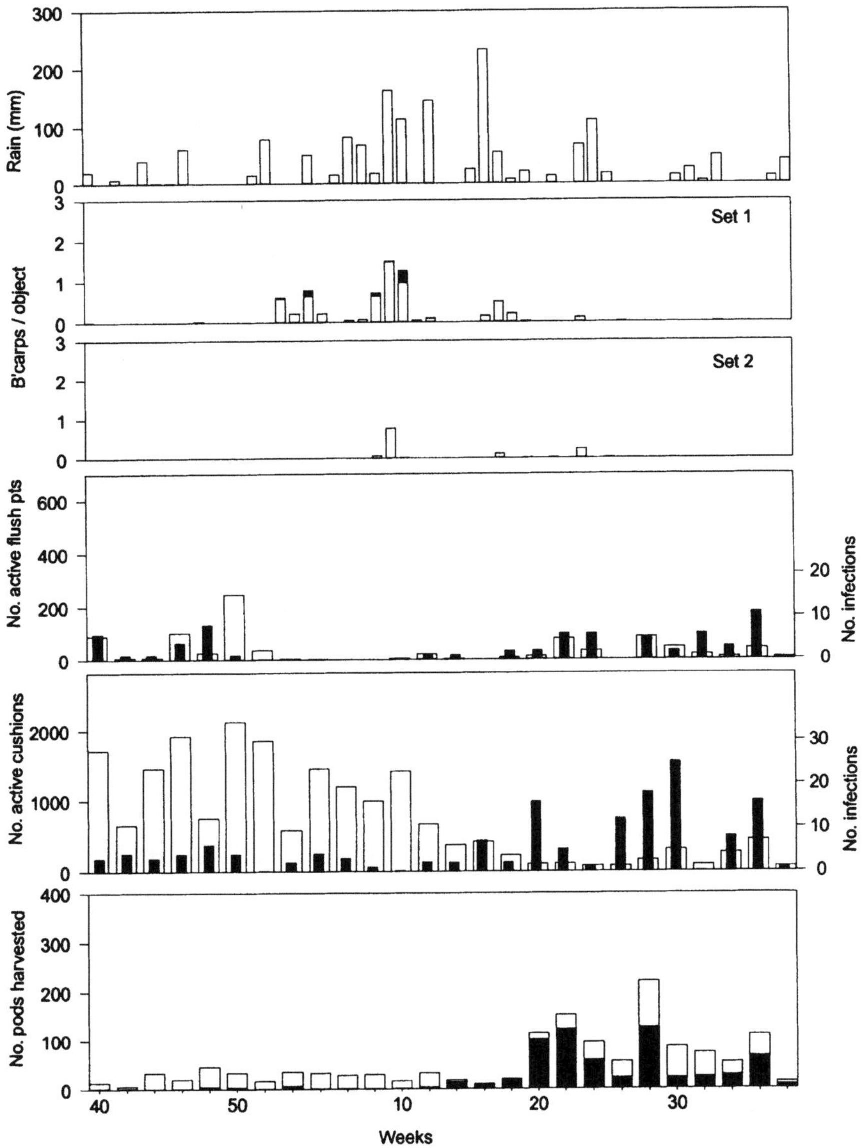

Figure 5.7. Rainfall, basidiocarp production[1], vegetative flushing and infection[2], flower cushion activity and infection[3], fruiting and fruit infection[4] in the Tomé Açu Site, Brazil during 1986/7.

[1] open columns, the most productive source type and closed columns, sum of the rest; [2] open columns, no. of healthy flushes and closed columns, no. of infections; [3] open columns, no. of active healthy cushions and closed columns, no. of infections; [4] closed columns, witches' broom infected, shaded columns, other diseases, and open columns, mature healthy fruit.

Disease symptoms mainly appeared on flower cushions when cushion activity was in decline or at its lowest level. Most diseased cushions were recorded between weeks 15 and 38 in the two years of the experiment, which could be attributed to infections arising during the main period of basidiocarp production (week 2–22). However, symptoms continued to be produced until week 6 of 1987, showing that extended incubation periods were common. Furthermore, cushion disease observed in the first 20 weeks of the IWBP experiment in 1985/6 must have been derived from infections during the previous rainy season. Incidence of cushion disease was not related statistically to basidiocarp numbers in either year; no significant regression analyses were found allowing for a range of incubation periods.

Flower brooms represented 45% of all disease symptoms in year 1. Vegetative cushion brooms formed on only 16% of all diseased cushions, either singly or associated with other classes of disease symptoms. In year 2, flower brooms (41%) were still the commonest symptom class, followed by vegetative cushion brooms (21.3%).

Fruiting and fruit infection

Fruit set in year 1 was not completely confined to one period, although most formed between weeks 6 and 28 of 1986. Pods became diseased from week 16 to week 36, at the same time as healthy pods were maturing. The majority of the pods developed within periods when basidiocarps were recorded on the source sets of brooms. Witches' broom disease accounted for 30.5% of pods from the 50 trees, and 20.7% from within the marked sample units. Due to the frequent and high rainfall in year 1, 7.2% and 15.6% of pods became diseased with *Phytophthora* pod rot on the larger and smaller samples, respectively.

Fruit set in year 2 occurred at similar times to year 1, although the number of pods set within the sample trees fell by 14%, and pod disease rose by 22%. Basidiocarp production on Set 2 basidiocarp sources was scarce, but diseased pods with witches' broom within the units and from 50 trees increased to 38.8% and 40.5%, respectively. Pod loss to *Phytophthora* pod rot in year 2 was 3.5%.

Comparative Analysis and Discussion

Basidiocarp production

Suspended vegetative brooms were the most productive and precocious source of basidiocarps in all the sites, and production on the top canopy vegetative brooms was highly correlated in all sites. The shortest periods of rain required before initiation of basidiocarp production on second year brooms were 1 week in Altamira and 6 weeks in Ouro Preto. Newly suspended brooms from Set 2 required 12 rainy weeks in Tomé Açu and 9 weeks in Altamira. The broom sets used in Ouro Preto were not necrotic when suspended, and 24 and 15 weeks with rain were required before the induction of basidiocarp production occurred in their first rainy season (Table 5.2).

Total production of basidiocarps was greatest on brooms in Altamira followed by Tomé Açu and Ouro Preto. The low basidiocarp index recorded in Ouro Preto confirmed previous observations (Andebrhan, 1985*a*). Broom sets (Set 1) in their second year in all three sites produced appreciable quantities of basidiocarps. Duration of production was greatest in Altamira and similar in the other two sites.

TABLE 5.2
Basidiocarp production by the suspended vegetative brooms in relation to rainfall in the three Brazilian sites

Source set Site	Annual total b'carps	b'carp index/ object	Active weeks	Annual rain (mm)	Rainy weeks	Induction period[1]	Weeks[2] with rain
Set 1 (year 1)							
Altamira	1,589	8.0	21	2,178	47	9	9
Ouro Preto	239	1.2	9	1,715	43	24	24
Tomé Açu	1,422	7.1	18	2,768	42	22	15
Set 1 (year 2)							
Altamira	721	3.6	25	1,857	47	1	1
Ouro Preto	420	2.1	20	1,897	35	9	6
Tomé Açu	1,208	6.0	17	1,630	27	15	15
Set 2							
Altamira	1,579	7.9	30	1,857	47	6	6
Ouro Preto	557	2.8	15	1,897	35	20	15
Tomé Açu	91	0.4	8	1,630	27	28	12

[1] the number of weeks before the first basidiocarp was recorded after onset of rains; [2] the number of weeks with rain required before the first basidiocarp was recorded.

Vegetative brooms and diseased pods on the ground hardly produced any basidiocarps in Tomé Açu and Ouro Preto, but ground brooms were the second most productive source in Altamira. These differences may have been related to the variation in the amount of time taken for objects on the ground to be covered by leaf-litter. The experiment started in November in Tomé Açu and Altamira, while the peaks of leaf-fall were usually in September and October (Scerne & Santos, 1987). Peaks of leaf-fall in Ouro Preto were in August and September, and the experiment was installed by the second week of August. Decomposition of the brooms would have been more rapid where they became most rapidly covered in leaf litter. Basidiocarp production on diseased pods was sporadic and sparse in all the sites and years.

Rainfall was the most important climate factor affecting basidiocarp production. The amount of rain itself can be critical, but its frequency, timing and duration are known to affect timing and intensity of production. Basidiocarp counts were taken only once each week, so it was decided that analysis of the relationship with rain should consider the previous week rather than the rain in the same week. The weeks of intense basidiocarp production on suspended vegetative brooms had between 51 and 68 mm rain in the previous week, and weeks with any basidiocarps were preceded by weeks with 51 to 109 mm. The most intense production occurred after moderate amounts of rain, not generally after extremes in either sense.

Minimum temperatures were above 20°C and maximum temperatures were below 32°C at times of peak basidiocarp production in all sites. Ouro Preto had the smallest range of extreme temperatures during the sporulation period. It was not possible to establish any relationship between temperature and basidiocarp production between the sites and years.

Flushing and Vegetative Infection

Before the start of the experiments, all diseased tissues were removed and the trees given a structural pruning. It is known that bud activity can be stimulated by pruning, and periods of shoot growth (flushes) were observed in all the sites 2–4 weeks after the pre–experiment removal.

The timing of shoot growth was not similar in the two experimental years either in Ouro Preto or in Tomé Açu. There were between 4 and 6 flushes in each year at all three sites, only some of which coincided with periods of basidiocarp production. Most disease symptoms appeared 4–6 weeks after each of the flushes exposed to inoculum. The lower shoot activity index (Table 5.3) in all sites in year 2 than in year 1 was attributed mainly to the lack of pruning at the start of year 2. Altamira had the highest shoot activity index and Tomé Açu the least activity in both years. The low activity index in Tomé Açu was attributed to the dense overhead shade and soil with low natural fertility. The new leaf index (see Table 5.3) in year 2 was slightly lower than that in year 1 and it appeared that structural pruning affected shoot activity indices more than the new leaf indices (see Table 5.3).

Brooms were removed from trees in the IWBP sites at the beginning of year 1, and vegetative disease indices were comparatively low in all three sites in the first year (Table 5.3). Disease indices in all sites rose appreciably in the second experimental year. Altamira was the site with the largest quantity of basidiocarps on sample sets and the highest shoot activity index in both years, and yet had the lowest disease indices of the three sites. The higher disease index in Tomé Açu occurred despite lower shoot activity indices and abnormally low numbers of basidiocarps produced on the new brooms. It is possible that higher disease indices in Tomé Açu were related to the greater shade–tree density at that site. Data on types of disease symptom showed that bud infections predominated over non–bud infections in every site and year, except for year 1 in Tomé Açu.

TABLE 5.3
Comparison of shoot activity and vegetative disease in the Brazilian sites

Year	Site	Shoot activity[1] index	New leaf[2] index	Vegetative disease[3] index
1	Altamira	4.9	3.8	1.2
	Ouro Preto	2.7	3.5	2.0
	Tomé Açu	2.4	3.5	5.1
2	Altamira	3.7	3.0	2.3
	Ouro Preto	1.9	3.3	2.9
	Tomé Açu	0.8	3.2	7.7

[1] accumulated total of active points/number of terminal buds; [2] accumulated total of new leaves/accumulated total of active points; [3] number of diseased points/number of terminal buds (x 100).

Cushion Activity and Infection

Some flowering occurred throughout the two experimental years in all three sites. There was a general annual cycle of flowering intensity, with the trough coincident with the time of greatest fruit load in May/June (weeks 20 to 30). Flowering was most intense at, and just after, the time of greatest fruit set in October to December. Although genotypes and tree ages were similar in all sites, the number of active flower cushion observed in the sample branch units varied between the sites. Tomé Açu and Ouro Preto had the lowest and highest number of cushions, respectively, in both years. Such difference were also reflected in the cushion activity indices (Table 5.4).

The cushion activity index in year 2 decreased in all three sites by 10–12%, despite an increase in the total number of cushion meristems in year 2 (Table 5.4). Reduced activity could have been directly related to the lack of structural pruning, but it was more likely due to the reduction in light reaching the branches by year 2 due to self-shading.

It was possible to derive indirectly an estimate for the period of disease incubation in flower cushions. In all the sites and in the two years, the periods of most intense basidiocarp production preceded by 6–10 weeks the main period of appearance of disease symptoms on cushions. Many symptoms appeared several weeks after the principal period of emergence, indicating a skewed distribution for the incubation period.

The cushion activity index was not related to the cushion disease index for the six site/year combinations in Brazil. Tomé Açu gave the lowest cushion activity index in both years, and yet the cushion disease index there was the highest in the second year (Table 5.4). Ouro Preto in year 1 had the highest activity index and the lowest disease index. The number of cushions active during the inoculum period was in itself unlikely to have been the factor limiting the incidence of disease on cushions. However, it is not clear from these data what factors determined the cushion disease index.

TABLE 5.4
Comparison of cushion activity and disease in the Brazilian sites

Year	Site	Total no. cushion meristems	Cushion activity index[1]	Cushion disease index[2]
Year 1	Altamira	2,733	16.3	0.7
	Ouro Preto	2,883	18.2	0.3
	Tomé Açu	2,181	9.5	0.6
Year 2	Altamira	2,864	14.7	0.3
	Ouro Preto	3,219	15.8	0.7
	Tomé Açu	2,543	8.1	1.3

[1] cumulative no. of active cushions / total no. of cushion meristems; [2] total number of diseased cushions / total no. of cushion meristems (x 100).

The predominant symptoms of infection in all the sites and years were flower brooms followed by cushions with vegetative brooms. In all the sites, presence of chirimoyas was sporadic.

Fruiting and Fruit Infection

In all the sites, diseased pods were harvested 8 weeks earlier than healthy ripe pods which were set during the same week. On average, healthy pods needed 22 weeks to mature. The number of fruits set on sample units was low and for analysis of cropping and infection patterns, pod harvests from the 50 trees were used. Trees at Ouro Preto produced the most pods and trees at Tomé Açu the least (Table 5.5). Mean pod losses (50 trees) were 11, 36.9 and 30.5% in Altamira, Ouro Preto and Tomé Açu, respectively.

Most fruit set at Ouro Preto and Tomé Açu occurred during periods of basidiocarp production. However, at Altamira substantial numbers of pods were formed outside the basidiocarp production periods. This, in part, explained the low pod losses recorded in both years. The total number of pods and the percentage lost to witches' broom increased in year 2 at Ouro Preto and Tomé Açu, but did not do so at Altamira. In all the sites, there were high bearing trees with very low pod losses, which implied that some trees had a degree of disease resistance.

With our present understanding of *C. perniciosa* biology in relation to climate, the environmental factors monitored did not vary enough to explain the large variation in witches' broom incidence on pods from site to site. When compared to other sites, Altamira had the greatest basidiocarp production in both years, but had the lowest pod losses. Timing of pod production in Altamira in relation to sporulation could partly explain the lower pod losses. Pod losses to other causes such as *Phytophthora* pod rot were negligible. It is also possible that variation in virulence of *C. perniciosa* could account for the lower incidence at Altamira.

TABLE 5.5
Numbers of all types of pods and the fraction diseased on 50 sample trees and 50 sample branch units in the Brazilian sites

Year	Site	50 whole trees		50 sample units (on 10 trees)	
		Number of pods	% witches' broom	Number of pods	% witches' broom
Year 1	Altamira	3,578	11.0	254	11.4
	Ouro Preto	5,017	36.9	239	23.8
	Tomé Açu	1,139	30.5	135	20.7
Year 2	Altamira	3,543	11.0	200	7.5
	Ouro Preto	5,443	53.4	183	27.9
	Tomé Açu	1,388	40.5	116	10.3

CONCLUSIONS

1. The total amount of rain varied considerably between years and sites. Tomé Açu was the wettest in year 1 and the driest in year 2. The timing of the rainy season did not change appreciably between years at each site. The average air temperature was 2–3°C higher in Tomé Açu than the other two sites, over the long term and two years of the experiment.

2. Basidiocarps were first observed on suspended vegetative brooms in all sites. Brooms at Altamira needed the least and brooms at Ouro Preto the largest number of rainy weeks before initiation of basidiocarp production. The main period of basidiocarp production in Altamira and Tomé Açu occurred between January and mid-June, and in Ouro Preto between October and April. Of the aerial sources, the suspended vegetative brooms were the most productive followed by the cushion brooms. Vegetative and cushion brooms at Altamira were the most productive. Basidiocarps were produced on ground brooms at the same time as on the canopy brooms, and their effect could not be separated.

3. Shoot activity was most intense before the start and near the end of the rainy seasons at all the sites, and it was either negligible or reduced in the middle of the rainy season. There were 5–6 peaks of shoot activity, only 1 or 2 of which coincided with the main period of basidiocarp production. Appearance of disease symptoms on shoots resulting from infections during the rainy season was extremely variable, and no standard incubation period could be calculated. Bud infections were predominant.

4. Flower cushions were active usually throughout the year. The number of infection courts was not a factor limiting witches' broom infections and most cushion disease could be attributed to infections that arose during the period of maximum basidiocarp production. The mean incubation period was 8 weeks, although the distribution was skewed. Flower brooms were predominant.

5. Most pods in Tomé Açu and Ouro Preto passed their first 12–14 weeks of development during the period when basidiocarps were present, and peak incidence of diseased pods slightly preceded peak incidence of healthy mature pods. Most pods matured between May and September and May and July in Tomé Açu and Ouro Preto, respectively. Pods matured in Altamira throughout the year.

6. From this study, it was confirmed that a cocoa field with heavier shade (Tomé Açu) had less intense shoot activity with fewer new growing points, less cushion activity and less pod production. Although disease indices on shoots were higher in the shaded plot, this was not necessarily the case for cushion and pod indices. The site completely without shade trees (Altamira) was severely attacked by pests such as *Monalonion* sp., which contributed to the increase of new growing points during the rainy period when inoculum was available.

7. In cocoa fields such as that in the Altamira site with flat topography, clay soil with good water retention and high natural soil fertility, a significant number of pods can set during the periods of low rainfall and escape periods of basidiocarp production. Annual pod losses are reduced as a result of such disease escape. Particular trees with escape characteristics were observed in Altamira, and such individuals could be considered in a regional breeding programme to reduce pod losses (see Chapter 16).

8. The minimum incubation period for cushion infection in Brazil was estimated at 6–8 weeks, although it was longer than this in many cases. Direct observation of basidiocarp sources is not essential to monitor patterns in basidiocarp production, as the incidence of new cushion symptoms would give a strong indication as to the timing of peaks of basidiocarp production. This technique would be particularly useful where there is a new outbreak of the disease, as in Bahia State, Brazil.

9. Fewer pods were set in the marked branch units in year 2 than year 1 at each site, but the total number of pods produced on the 50 trees increased in Ouro Preto and Tomé Açu between the years. The trees were pruned at the start of year 1, and it is possible that overall yield was stimulated by removal of parts of the canopy. It is also possible that sanitation pruning could have positive effects on production from cocoa trees and the possible relationship should be studied further.

MANAGEMENT IMPLICATIONS

1. Management of the disease should aim at eliminating the sources of inoculum. Basidocarps of *C. perniciosa* in the Brazilian IWBP sites were principally derived from the brooms in the canopy, with little if any derived from brooms on the ground. All such aerial sources of basidiocarps should be removed (by phytosanitation) before the start of the main rainy season or at the start of the dry period, as rain is the principal factor controlling basidiocarp production on necrotic diseased tissues.

   1.1 In areas of high disease incidence: phytosanitation needs to be completed during the months of August/September, after the main pod harvest and when most disease symptoms have appeared. A secondary sanitation (repass) is necessary during October to December, to remove those diseased tissues appearing after a delayed incubation, as well as any missed brooms.

   1.2 In areas of low disease incidence: phytosanitation should be done at the start of the dry period, and irregularly at any period of the year in association with other agronomic practices (pod–harvests, chupon removals, etc.).

2. Diseased material removed from the trees may either be left on ground within the plantation or removed from it. However, the practice of removal of the diseased tissues from the plantation alone increases the cost of the sanitation control by 50% (Almeida & Andebrhan, 1988).

   2.1 In areas of high disease incidence or with overhead shade, the diseased material removed can be left on the ground within the plantation, if it is cut up and can be easily covered by naturally falling leaf debris.

   2.2 In areas of low disease incidence, without overhead shade or after a severe pruning, ground brooms can produce significant numbers of basidiocarps. In such situations, every effort should be made to cover the diseased material with leaf litter.

3. Most pods in the Brazilian IWBP sites formed on branches less than 4 m from the ground. The cost and efficiency of phytosanitation is related to the ease of access to diseased plant parts, which is in turn dependent on tree height. An experiment in Altamira has shown that structural pruning over 3 years to keep tree height to less than 3.5 m did not affect pod production, compared to trees left to grow taller (Andebrhan, 1985*c*), and trees should be restricted to that height where possible.

4. The use of fungicides as a complement to phytosanitation should be cost–effective and ecologically acceptable. The application of any chemicals should take into consideration the phenology of the infection courts and the dynamics of inoculum production in each site.

5. In the short term, agronomic practices which would increase fruit set, thus compensating for diseased pods, should be considered. In the long term, breeding for durable resistance and selecting trees with disease escape should be given the highest priority.

Chapter 6

COMPARATIVE EPIDEMIOLOGY STUDY : COLOMBIA CALDAS AND LLANOS ORIENTALES

F. Aranzazu H. & P. Buriticá C.

INTRODUCTION

Importance of Cocoa and Witches' Broom in Colombia

Colombia is a country with a long tradition of cocoa growing, and chocolate is a basic part of the national diet. Production in 1989 was estimated to be 51,000 t from a harvested area of 103,000 ha within the total of 120,000 ha of cocoa planted. Cocoa producing regions are distributed throughout the country according to the varied physical geography and the climatic zones, and official figures for production (FEDECAFE) show 26 departments in the country as producers of cocoa. Colombia has some of the highest altitude commercial plantations in the world at 1,300 m a.s.l., and has a significant area of artificially irrigated crop. The only major concentration of cocoa is in the central department of Santander, which produces about a third of the national total. The average yield of 500 kg/ha is low principally due to the fact that almost 50% of the cultivated area consists of traditional plantations of old cocoa, and because of attack by such limiting diseases as moniliasis (*Moniliophthora roreri*) and witches' broom (*Crinipellis perniciosa*).

Witches' broom was recorded in Colombia for the first time in Tumaco in 1928. Only in 1962 did the disease begin to spread to other cocoa growing areas of the country, such as the neighbouring Departments of Huila and Valle in central Colombia, Antioquia in the northern coastal plains and Los Llanos Orientales (the eastern plains). In Antioquia (Urabá region) at the end of the 1970s, the disease was not controlled and it led to the destruction of 5,000 ha of hybrid cocoa. Losses were in many cases in excess of 60% of the harvest. The disease is also apparently favoured in Los Llanos by higher rainfall and relative humidity, and is known to cause losses in excess of 40% of pods.

The disease was first recorded in 1973 in the cocoa growing zone of the Department of Caldas, though it now causes a fruit loss of less than 10% and is not a major factor limiting production. Finally, witches' broom was detected in 1989 in the Department of Santander in the middle Magdalena valley, leaving only the extreme north of the country unaffected by the disease.

Disease Management in Cocoa: Comparative epidemiology of witches' broom. Edited by S.A. Rudgard, A.C. Maddison and T. Andebrhan. Published in 1993 by Chapman & Hall, London. ISBN 0 412 58190 6

Witches' Broom Research in Colombia

Only at the end of the 1970s did a series of investigations begin in Colombia which have led to relatively successful disease control. Many findings have already been described in Chapter 2. A few points relating to the previous knowledge of witches' broom in Colombia will be mentioned here. The two principal series of experiments were based in Urabá and Los Llanos, where disease incidence was highest. A complete review of 12 years work in Los Llanos has been compiled by Tovar (1991).

The main production of basidiocarps in Urabá was known to occur in October–November (Aranzazu, 1982*b*). In Los Llanos, there was great variation between years, but significant quantities of basidiocarps were produced between April and November (Tovar, 1991). In every case basidiocarp production has shown a close relationship with monthly rainfall (Aranzazu, 1982*b*; Cifuentes *et al.*, 1982*b*; Tovar, 1991). Flowering and the principal vegetative flushes of cocoa trees in Los Llanos were also known to occur during the first half of the year (Cifuentes *et al.*, 1982*a*; Tovar, 1991). Fruit set and development and some isolated shoot flushes took place in the second half of the year.

Most brooms were found to form in Urabá and Los Llanos at the end of the rainy season and beginning of the dry period January–February, and the least from May to June. The highest fruit losses to witches' broom always occurred between August and September (Aranzazu, 1982*a*; Cifuentes *et al.*, 1982*b*; Tovar, 1991). In contrast, most brooms formed in July–August in the marginal lower coffee zone of Caldas, with a second peak in February and a minimum in May (Aranzazu, 1986).

Current Practices for Disease Management

Phytosanitary pruning once or twice a year has been shown to be most efficient and economic (Aranzazu, 1982*a*, 1986; Cifuentes *et al.*, 1982*a*; Rondon, 1986; Tover 1991). Tovar (1991) also provided evidence that brooms on the ground presented a small enough risk for pod infection that pruned brooms could be left in the plantation. Control of the disease in severely affected plantations has been achieved by renovation of the canopy and reduction of the height of the trees (Cubillos & Grisales, 1985). Investigations into chemical and genetic control have not been extensive and no positive results have been reported.

METHODS

The experiment was carried out in two geographically and ecologically very different cocoa-growing areas of Colombia.

Montelindo, Manizales

One site was established in the Department of Caldas at the Montelindo Farm of the University of Caldas, situated at 1,020 m a.s.l. in undulating terrain. The cocoa plantation in which this study took place was 25 years old, had an area of 7 ha and was isolated by a of more than 500 m from other plantations. The 10 trees of the clone SC–6 used for detailed recording were surrounded by 32 trees of the same clone used for further harvest data. All developing fruits were removed from the 10 sample trees before beginning the study. As well as the information agreed in the protocol, the number of weeks between the appearance of vegetative brooms on the tree and the beginning of basidiocarp production was recorded. The experiment lasted for two years, from week 14 of 1986 to week 13 of 1988.

Los Llanos

The second site was established in Los Llanos Orientales of Colombia in the municipality of Granada at La Cabaña Farm, situated 300 m a.s.l. The experimental area was part of a cocoa plantation of 100 ha composed of a mixture of 8 year old hybrids under moderate and well distributed shade. The 10 trees used for detailed recording were surrounded by a total of 39 other trees. The site was flooded during part of 1986, and some social problems in the middle of 1988 interrupted collection of data. For this reason it was decided to analyse only the data for 1987 (12 months).

Recording of numbers of basidiocarps in the two sites departed from the Protocol in two ways. The different sources were examined weekly on a particular day, not necessarily after rain, so the basidiocarps included in the count were not always turgid. Also the leaf-litter covering the sources placed on the ground was removed each week, allowing greater circulation of air and drying of the brooms/pods. In Los Llanos, fruits were normally harvested every two weeks, but the analysis presented is monthly because some harvests were missed.

RESULTS AND DISCUSSION

Montelindo, Manizales

Climate

In Figure 6.1, average monthly rainfall is shown for Montelindo (Caldas), including 10 year means and separate data for the first and second years of the experiment. The zone has an average rainfall of about 2,000 mm well distributed through the year without any excessively dry months. The wettest months are April–May and October–November; January, February, July and August are the driest months. The mean annual temperature is 24°C, and the average annual relative humidity 78%. Precipitation for 1986 and 1987 was average, with the month of October being the wettest with around 350 mm rainfall.

Basidiocarp production

Set 1 suspended vegetative brooms placed half-way up and at the tops of the trees produced the greatest number of basidiocarps per object, followed by the cushion brooms (Table 6.1). Suspended vegetative brooms and cushion brooms were observed with basidiocarps in 8 weeks on average, and the other sources were active for about 4 weeks. Brooms which survived more than a year continued to produce basidiocarps, albeit less often. Set 2 diseased tissues had higher basidiocarp indices, and were active more often than Set 1 during the first

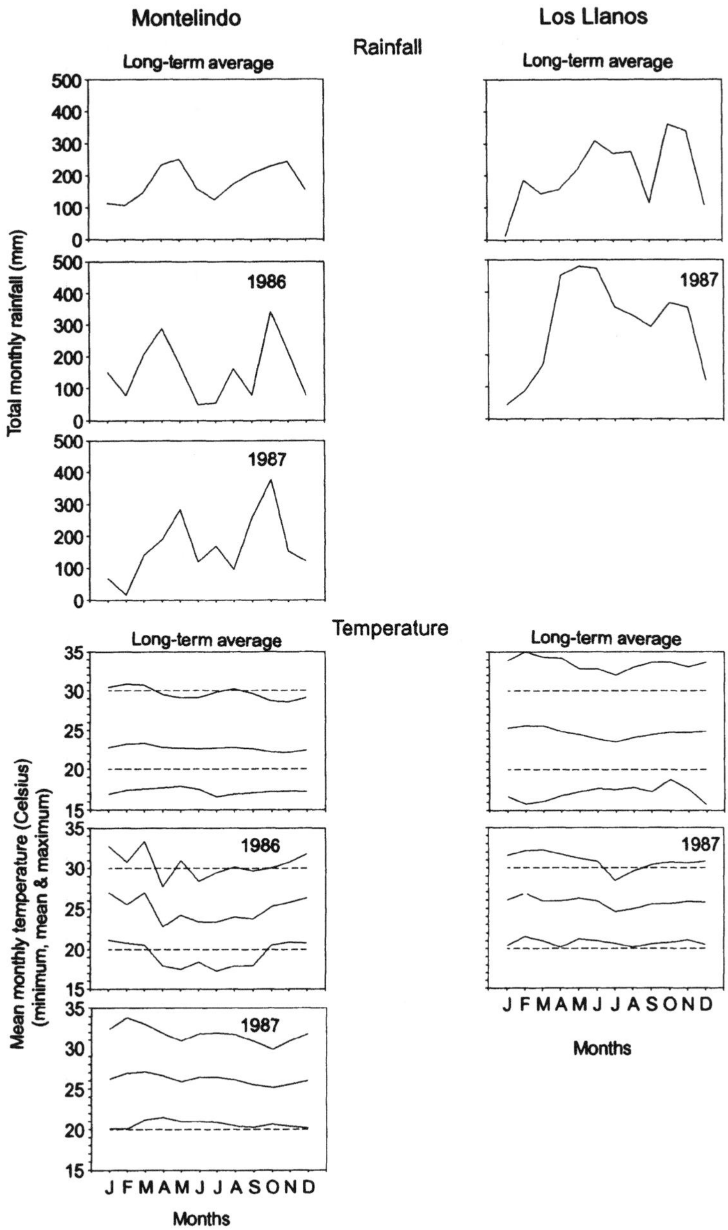

Figure 6.1. Rainfall and temperature data compared with long-term averages from the Montelindo and Llanos Sites, Colombia.

year, possibly because 1987 was somewhat wetter than 1986. In 1987, cushion brooms were more productive than vegetative brooms, followed in order of importance by pods on the ground, vegetative brooms on the ground and suspended fruits.

TABLE 6.1
Numbers of basidiocarps produced during the two years by the two principal sources, Montelindo, Manizales

	Vegetative brooms			Cushion brooms		
	Set 1 Year 1	Set 1 Year 2	Set 2	Set 1 Year 1	Set 1 Year 2	Set 2
Total b'carps	4,089	2,930	4,350	2,570	1,458	5,922
Mean b'carps/broom	20.4	14.7	22.0	12.9	7.3	29.0
Mean active weeks/broom	9.2	5.5	7.5	7.4	5.5	8.4
B'carps/broom/active week	2.2	2.7	2.9	1.7	2.7	3.5
% brooms active	98	99	96	83	97	97

An average of 6 weeks elapsed from the formation of brooms in the tree to the start of their necrosis. From necrosis to the start of sporulation a minimum of 20 weeks and an average of 36 weeks were necessary. Thus there was a period of 42 weeks before they started to produce basidiocarps.

There was no long period during which basidiocarp production was inhibited (Figures 6.2 and 6.3). Production always declined at the beginning and in the middle of each year, coinciding with weeks without rain, or weeks when rain occurred sporadically during the day. Such weather is characteristic of the variable dry season of the central coffee zone. The analysis of variance of the regression between numbers of basidiocarps and rainfall was highly significant in the two years, with correlation coefficients of $r = 0.56$ and 0.47. The relationships were also significant for the number of rainy days ($r = 0.55$ and 0.59), and for days with leaf wetness of more than 10 h duration after 22.00 h ($r = 0.50$ and 0.51). Following rain, sources placed on the ground produced basidiocarps more rapidly than the suspended sources, possibly due to a greater conservation of humidity at ground level.

Analysis of the behaviour of individual brooms showed that each broom had a different cycle of basidiocarp production. However, on occasions the majority of the brooms' cycles coincided with one another, giving weeks with a high percentage of active brooms and a high basidiocarp index. The weeks of most intense production on the suspended sources were weeks 17, 27 and 42 in 1986, and weeks 18, 32 and 44 in 1987. These weeks, with an average of approximately 50% of the brooms active, also had the greatest number of rainy days, and the greatest number of days with leaf wetness of more than 10h duration after 22.00 h.

Flushing and vegetative infection

The pattern of vegetative shoot growth in clone SC–6 was clear, with little variation in timing

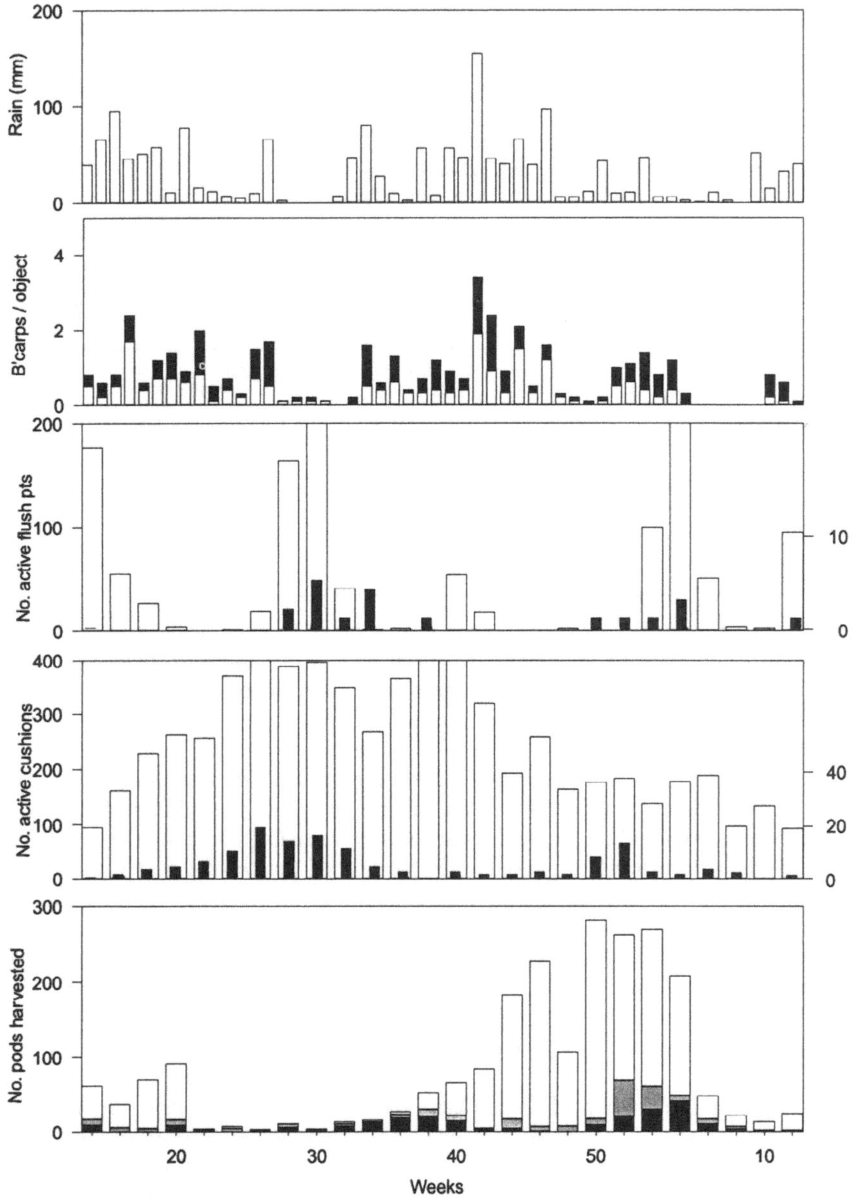

Figure 6.2. Rainfall, basidiocarp production[1], vegetative flushing and infection[2], flower cushion activity and infection[3], fruiting and fruit infection[4] in the Montelindo Site, Colombia in 1986/7.

[1] open columns, the most productive source type and closed columns, sum of the rest; [2] open columns, no. of healthy flushes and closed columns, no. of infections; [3] open columns, no. of active healthy cushions and closed columns, no. of infections; [4] closed columns, witches' broom infected, shaded columns, other diseases, and open columns, mature healthy fruit.

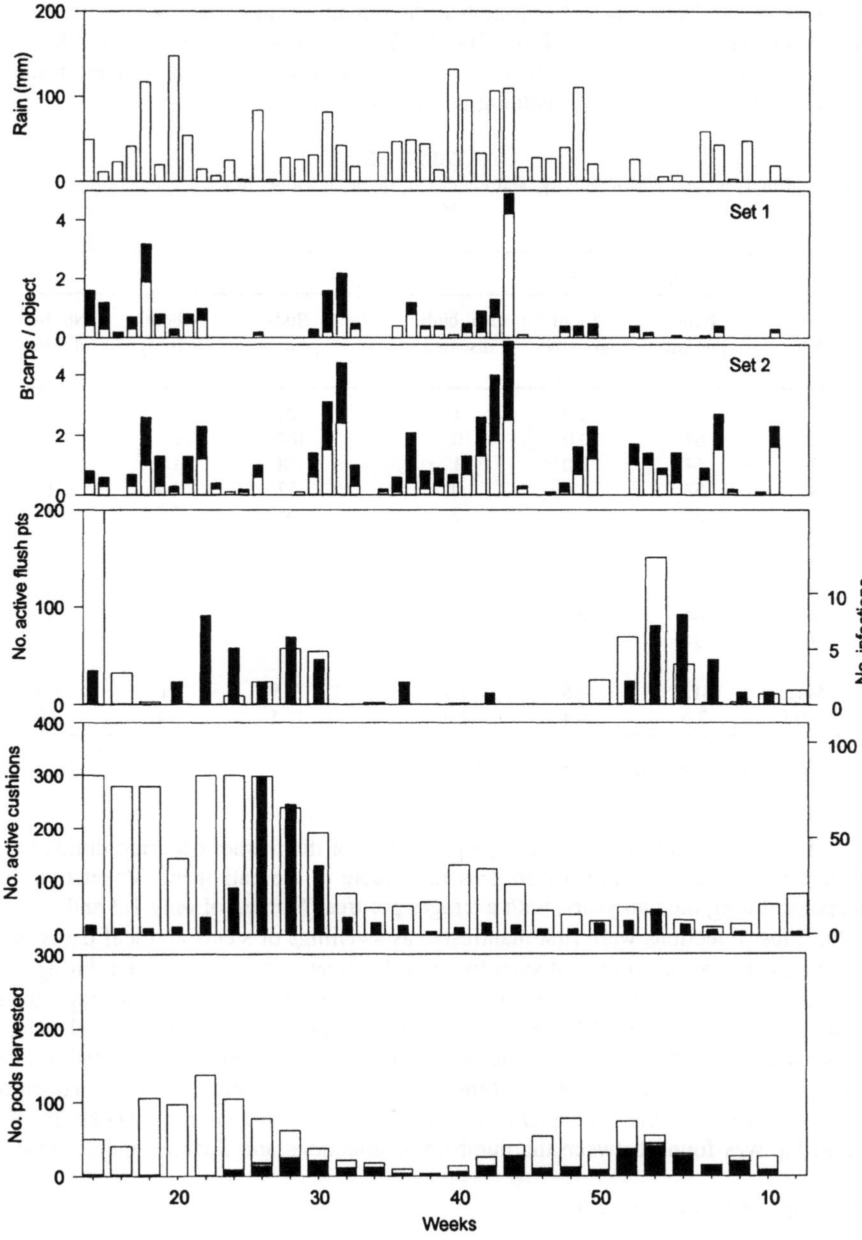

Figure 6.3. Rainfall, basidiocarp production[1], vegetative flushing and infection[2], flower cushion activity and infection[3], fruiting and fruit infection[4] in the Montelindo Site, Colombia in 1987/8.

[1] open columns, the most productive source type and closed columns, sum of the rest; [2] open columns, no. of healthy flushes and closed columns, no. of infections; [3] open columns, no. of active healthy cushions and closed columns, no. of infections; [4] closed columns, witches' broom infected, shaded columns, other diseases, and open columns, mature healthy fruit.

between trees. In both years, the clone had three main shoot flushes which occurred in the same weeks, namely: 2–4, 12–14 and 28–30. The shoot activity index was 2.6 in the first year and 1.0 in the second, possibly due to the excess foliage not being pruned in the second year which led to many buds remaining dormant (Table 6.2).

TABLE 6.2
Numbers of buds present, shoot activity, and number of diseased shoots in the 10 sample trees, Montelindo, Manizales

	Year 1 : 1986			Year 2 : 1987		
Tree No.	Buds present	Shoot activity index[1]	No. buds diseased	Buds present	Shoot activity index[1]	No. buds diseased
1	55	2.9	1	70	1.5	3
2	67	3.0	0	100	0.7	5
3	64	3.0	1	78	1.4	16
4	57	2.4	0	57	1.4	6
5	56	2.6	2	68	0.2	5
6	60	2.9	2	70	0.8	6
7	57	2.0	0	90	1.1	4
8	79	2.2	2	79	0.8	1
9	64	2.8	3	77	0.9	2
10	54	2.4	2	70	0.7	5
Mean	61.3	2.6	1.3	75.9	1.0	5.3
S.D.[2]	7.6	0.4	1.0	12.1	0.4	3.9

[1] sum of flush points/sum of terminal buds; [2] standard deviation.

It is important to note that a large proportion of the flushes occurred during the weeks with less rain, when conditions were less favourable to the pathogen. Vegetative infection was sparse during the two years, with averages per tree (5 units) of only 1.3 and 5.3 diseased shoots. Most infections were first manifested as swellings of stems and leaf bases, and gave the impression that infection and symptom development had both occurred during the same flush period. In a previous study with this clone, the minimum and average incubation periods in shoots were found to be 28 and 47 days respectively (Giraldo & Valencia, 1985). No correlation was found between the production of basidiocarps and vegetative infections occurring 7, 8 and 9 weeks later. Most diseased shoots were first recorded during the vegetative flushes of the first quarter (weeks 2–4) and third quarter (weeks 28–30). No relationship was found between the numbers of infections and flushing indices (Table 6.2).

Flowering and cushion infection

Clone SC–6 flowered continuously during the two years, with a long period of more abundant flowering during the second and third quarters of the year (Figures 6.2 and 6.3). The total of active cushions was greatly reduced (53%) in the second year (Table 6.3), though the timing of the maxima and minima stayed the same. The reduction was apparently due to excessive self–shading caused by the lack of pruning, the same phenomenon as occurred with vegetative shoots.

TABLE 6.3
Flower cushion activity and infection within the flowering units in the 10 trees for the two years, Montelindo, Manizales

Tree	Number of cushions			
	1986		1987	
	Active	Diseased	Active	Diseased
1	747	16	407	23
2	599	34	289	58
3	733	12	442	39
4	727	8	381	42
5	449	4	260	24
6	740	12	289	12
7	856	3	395	25
8	280	7	229	24
9	686	19	525	22
10	780	24	270	30
Total	6,597	139	3,487	299
Mean	659.7	13.9	348.7	29.9
S.D.[1]	164.3	9.2	90.5	12.4

[1] standard deviation.

New records of diseased cushions were most common during the third quarter of the two years between weeks 24 and 34, and slightly less so around the new year (weeks 50–4). Formation of new symptoms continued at other times but at a lower intensity. It was already known that the minimum incubation period for cushion disease in the clone SC–4 was 8 weeks (Mayorga, 1984). It was inferred that the principal period of cushion disease was due to infections from about week 17 or 18 in both years, when there was maximum sporulation, conditions of rain and humidity favourable to the fungus, and abundant flowering. The relatively fewer infections in weeks 50–4 could be related to the low intensity of flowering during the fourth quarter in weeks 42 and 44, despite intense sporulation and apparently favourable climatic conditions.

The linear regressions of numbers of cushion infections on basidiocarp numbers 8 or 9 weeks earlier were not significant, but there was some coincidence between cushion infection peaks and the periods of particularly high basidiocarp production (Figure 6.3). The average per tree of diseased cushions within the units during the first year was 13.5 ± 9.0 and increased to 29.4 ± 12.0 for the second, an increase of 48% (Table 6.3).

Fruiting and fruit infection

The initial period of fruit growth, which is their time of greatest susceptibility, coincided with the wettest periods, the second and fourth quarters of the year. During the first year there was only one rather long harvest season which lasted from weeks 42 to 4. No pods were harvested during the first quarter, because all fruits were removed from the 10 trees at the beginning of the trial due to the high incidence of moniliasis. The second year was more usual for the region, showing two definite harvest peaks, the first in May–June and the other in December–January.

The greatest incidence of witches' broom diseased fruits was recorded during the first and third quarters, approximately 12–14 weeks after the key periods of high to medium production of basidiocarps and high humidity. The analysis of variance for the linear regression of infected fruit on basidiocarp numbers 13 weeks earlier was significant at the 5% level, and had an r value of 0.51 ($y = 63.1 + 3.2\ x$). The large intercept indicated that many basidiocarps were present on the broom sets in some weeks in which no infection occurred.

The percentage of diseased pods with witches' broom was 12.0 in the first year (ranging from 7.7 to 22.2% in the ten trees) and 20.0% for the second (ranging from 11.0 to 57.1%). The figures for moniliasis were only 7.0% and 1.0% in the two years, due to the cultural control provided by harvesting frequently before the diseased pods could sporulate.

Comparisons were made between the number of basidiocarps per week on suspended vegetative brooms and the number of infections in fruits 14 weeks later, and in cushions and shoots, 8 weeks later. Once such allowances for incubation had been made, the timing of the peaks of infection in the different tissues indicated that there were periods when conditions were particularly suitable for infection. In certain cases, these coincidences of infection occurred during periods of intense basidiocarp production (week 18, 1987), but there were periods with many basidiocarps (week 17, 1986) which did not have corresponding peaks of infection.

Los Llanos

Climate

For Los Llanos (Figure 6.1), the average annual rainfall over the last 30 years has been 3,600 mm, but values have diminished to around 2,500 mm perhaps due to the cutting of forest in recent years. December, January and February are the driest months of the year; April, May, October and November the wettest. The long-term average temperature is 26°C with a maximum of 30°C and a minimum of 18°C; average annual relative humidity is 80%. For 1987, 2,547 mm of rain were recorded, and October and November were the wettest months with more than 300 mm.

Basidiocarp production

In Los Llanos the most productive sources of basidiocarps were cushion brooms, with an index of 19.0 basidiocarps/broom/year (Table 6.4). This productivity resulted from their high percentage activity (97%), together with a high average number of active weeks per broom (7.0). The second most active source of basidiocarp production was the brooms placed on the ground, with an index of 18.0 basidiocarps/broom/year, 95% of brooms active, and an average of 5 active weeks per broom. The remaining sources, the suspended vegetative brooms and diseased pods, and the pods on the ground produced a combined mean of 5.2 basidiocarps/object/year.

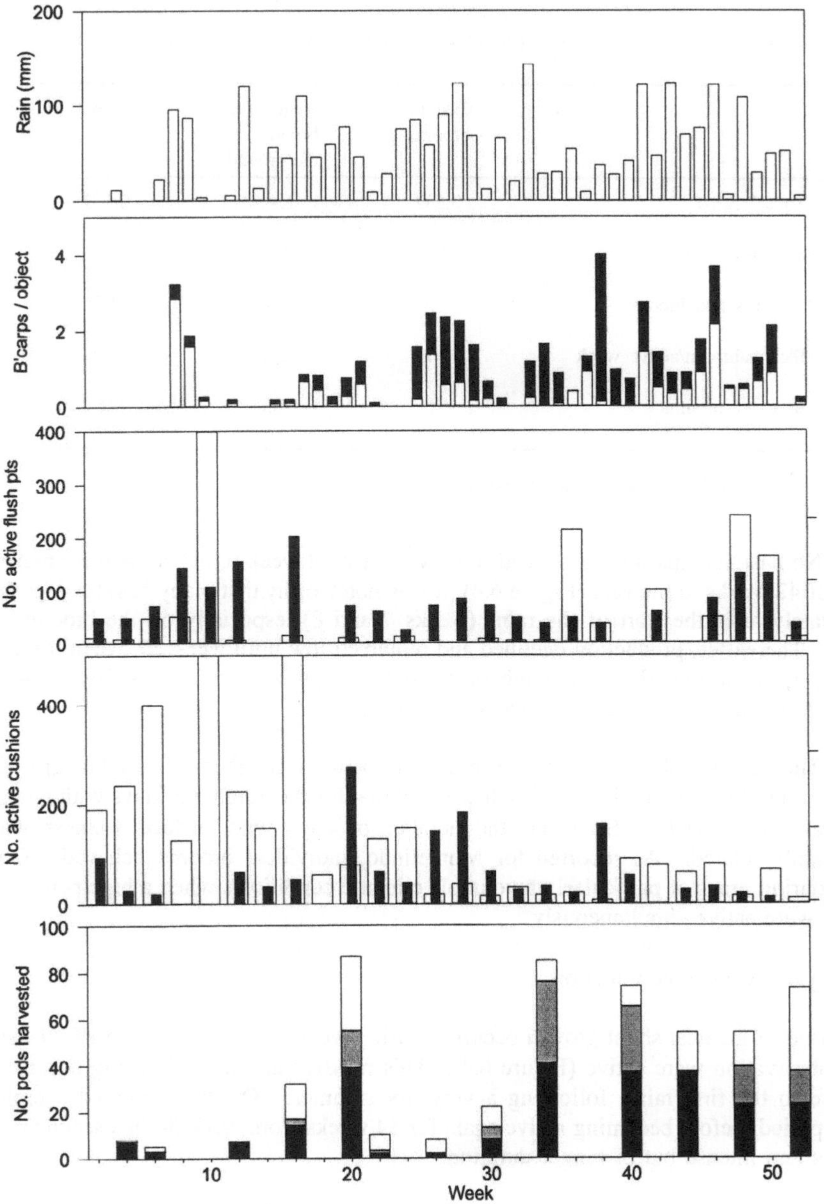

Figure 6.4. Rainfall, basidiocarp production[1], vegetative flushing and infection[2], flower cushion activity and infection[3], fruiting and fruit infection[4] in the Llanos Site, Colombia in 1987.

[1] open columns, the most productive source type and closed columns, sum of the rest; [2] open columns, no. of healthy flushes and closed columns, no. of infections; [3] open columns, no. of active healthy cushions and closed columns, no. of infections; [4] closed columns, witches' broom infected, shaded columns, other diseases, and open columns, mature healthy fruit.

TABLE 6.4
Comparison of the production of basidiocarps by the two principal sources, Los Llanos

	Cushion brooms	Vegetative brooms on the ground	Others[1]
Total b'carps	3,589	1,944	2,179
B'carps/broom	19.0	19.4	6.2
Active weeks/broom	7.0	5.0	1.2
B'carps/broom/active week	2.7	3.9	5.2
% active brooms	97	95	46

[1] suspended brooms and pods, and pods on the ground.

No basidiocarps were produced from week 1 until week 6, but some were present for a total of 42 weeks in the year (Figure 6.4). It was noteworthy that many basidiocarps formed immediately after the start of the rains (weeks 7 and 8), especially on the brooms on the ground. Thereafter, production declined and remained low until week 25, when it increased again to give particularly high numbers of basidiocarps between weeks 26 and 45. No basidiocarp readings were taken in weeks 12 and 13.

The analysis of variance of the regression between numbers of basidiocarps and the amount of rainfall was significant ($P < 0.05$), and that for the number of days with rain highly significant ($P < 0.01$). However, the number of days with surface wetness was not significantly related. As reported for Montelindo, individual brooms behaved differently except during times of particularly favourable climatic conditions, when a high percentage of brooms were active simultaneously.

Flushing and vegetative infection

The period of greatest shoot growth occurred during weeks 9 and 10 when 84% of the buds under observation were active (Figure 6.4). This massive and rapid flushing occurred as a response to the first rains, following a very dry summer. The trees entered a prolonged resting period, before becoming active again for 14 weeks from week 34 in a secondary flush that was less intense but of longer duration.

The highest number of infections appeared early in 1987, very possibly as a result of the prolonged flushing in late in 1986 (data not presented), and late in 1987 when a long flush of shoot growth coincided with the main period of basidiocarp production and conditions favourable for infection (weeks 37–45). Most vegetative infections were initially manifested as swellings and cankers. The mean flush index was 1.6 buds per tree per year, with a mean of 15.7 infections per tree (Table 6.5). There was no relationship between the number of infections and flushing index for the 10 trees.

Flowering and cushion infection

Flowering, in terms of the number of active cushions, was highest between January and April, and reduced during the rest of the year (Figure 6.4). The mean number of diseased cushions was 17.3 ± 15.9 per tree (Table 6.5) and most appeared between weeks 20 and 30 when few vegetative symptoms were manifested. These diseased cushions could have become infected in the period of intense cushion activity in weeks 16 and 17. Few basidiocarps were formed at that time, but both weeks were exceptionally wet (*ca.* 150 mm of rain), each with 3 days with surface wetness of more than 10 h after 22.00 h and about 150 mm of rain.

TABLE 6.5
Shoot activity, flowering, and disease in the 10 sample trees, Los Llanos

Tree	Number of vegetative buds			Number of cushions	
	Present	Shoot activity index[1]	Diseased	Active	Diseased
1	225	1.2	33	491	5
2	104	1.2	8	499	12
3	96	1.5	14	376	39
4	112	1.8	10	193	5
5	140	1.8	1	169	4
6	80	1.7	22	174	4
7	127	2.2	25	643	36
8	112	2.0	15	723	23
9	131	1.2	21	449	44
10	70	2.2	8	63	1
Mean	119.7	1.66	15.7	378.3	17.3
S.D.	43.0	0.37	9.6	221.2	16.7

[1] sum of active buds/sum of terminal buds.

Fruiting and fruit infection

There were, as usual for that region, two harvest periods in 1987, the first minor one in May and the second, more abundant, in November–December. Pod losses to witches' broom over the whole year were 54% and for *Phytophthora* pod rot were 17%. The average total of pods per tree was 16, very low for the age and type of the material. Accurate interpretation of the incidence of disease was difficult, owing to the four week interval between fruit data. However, some general associations between numbers of diseased pods and other factors can be made.

The initial high incidence of necrotic pods recorded in week 19 may have been related to infections in weeks 7 and 8, allowing for incubation. Large numbers of basidiocarps were noted at that time, and it has also been stated above that there was abundant rain in those weeks. The bulk of diseased pods became necrotic after week 33, and they must have become infected from around week 23. The main period of basidiocarp production was from week 25 to week 48 as already stated, when environmental conditions were favourable for infection, and developing fruits were susceptible.

CONCLUSIONS

1. The brooms with the highest basidiocarp production index and highest average of active weeks in Montelindo were the vegetative brooms suspended in the tree, and the second highest indices were obtained for cushion brooms. The principal source in Los Llanos was diseased cushions followed by diseased tissues left on the ground.

2. In Montelindo, there were no definite periods when sporulation was inhibited, because there were no definite dry periods. However, January–March in both years was the period of least basidiocarp production and least rainfall. Inhibition of basidiocarps in Los Llanos occurred during the first months of the year because of lack of rainfall. The time of greatest sporulation was recorded between weeks 26 to 48, which was also the wettest period.

3. In both sites, it was noted that on various occasions weekly basidiocarp production did not bear a very close relation to the quantity of rain. Individual brooms or groups of brooms became temporarily inactive after a period of active production. Basidiocarp numbers and broom activity were also reduced on brooms in their second active year.

4. In the two sites, basidiocarps formed throughout the wet season. However, the weeks with the greatest number of basidiocarps and greater broom activity occurred during the wettest period, both in the amount of rain and the number of days with rain and with surface wetness for more than 10 h duration after 22.00 h.

5. The physiological state of the trees, especially with respect to fruit development and flowering, was critical to interpreting incidence of infection. Incubation periods are known to be 10–14 weeks for fruits, and 6–8 weeks for cushions. Allowing for incubation from appearance of symptoms, infection was highest in the same critical periods for the two courts. However, numbers of basidiocarps alone did not explain the changes in amounts of infection. Susceptible host tissues were required during the appropriate climatic conditions.

6. The physiological behaviour of the host material, in which the genetic makeup and the climate are involved, seemed to be the principal factor in the development of witches' broom disease. Amounts of disease (symptoms) detected some weeks after infection, appeared to depend more on the physiological status of the trees than on the quantity of inoculum. The latter was generally present throughout wet periods, and sporadically during the changeable dry periods. Clone SC–6 at Montelindo proved to be highly susceptible to flower cushion infection through its constant flowering. However, it escaped vegetative infection as its flushes, intense but short, tended to coincide with dry weeks or weeks of low rainfall and low inoculum levels. At Los Llanos the hybrid material showed short lived but frequent flushes, and reduced but persistent flowering during the rainy season. Infection was more constant and continuous, albeit in lower quantities, than in Montelindo. This behaviour illustrates the advantage of these hybrids in respect of their resistance to witches' broom, relative to some local materials.

MANAGEMENT IMPLICATIONS

Management of witches' broom disease should be attempted only once the climatic cycles and the phenology the cocoa tree are known. These factors determine the most suitable time for man to intervene in order to break the disease cycle. The most vulnerable point appears to be production of basidiocarps, and the critical management method remains the removal of sources of inoculum within the planting. The key to success is the correct timing of this phytosanitation.

1. In plantations where witches' broom is already present, the most appropriate time to start control through the removal of diseased material from the trees should be at the middle and end of the principal dry periods. There is the least amount of inoculum in the environment, and most of the diseased tissues are necrotic and easy to see. According to the above results, phytosanitary and structural pruning in the central coffee zone around Manizales should be completed together between February and March. In Los Llanos, the recommended practice should be to execute the phytosanitary and structural pruning in February and March, but to follow up with a secondary phytosanitary pruning in June–July.

2. Pruned brooms wherever possible should not be left exposed on the ground, but leaf litter should be placed on them or allowed to accumulate so that they decompose rapidly. It should be pointed out that in Colombia, although there is no concrete evidence of the importance of the sources of inoculum found or placed on the ground, satisfactory control of the disease has been achieved even when pruned material is not removed from the plantation.

3. Management of the architecture of the tree in developing plantations, by reducing the height or controlling it to less than 4 m, should be considered as a basic and essential recommendation within control strategy.

4. Application of fungicides may become a feasible practice. Knowledge from each zone of the weeks with the highest ambient humidity would be fundamental to the protection of the developing fruits with fungicides in productive, managed plantations.

Chapter 7

COMPARATIVE EPIDEMIOLOGY STUDY: ECUADOR

A.C. Maddison, G. Macias, C. Moreira & J. Aragundi

INTRODUCTION

Importance of Cocoa and Witches' Broom in Ecuador

The area of cocoa presently cultivated in Ecuador is estimated at approximately 250,000 ha. There are about 55,000 families involved in its cultivation, which is usually in the range of 60,000–90,000 t per year. Cocoa is the fourth most important export from Ecuador, following oil, shrimps, and bananas. In the principal producing areas to the west of the Andes cocoa, like coffee, is grown typically as part of a extensive, but irregular agroforestry system in an area which has a high population growth and is susceptible to urban migration.

Witches' broom and moniliasis are the main factors limiting production, and together regularly cause more than 40%, and sometimes up to 80% pod loss. Precise determination of the relative importance of the two diseases is difficult for various reasons, not least because of the problem of distinguishing them on a proportion of pods at the moment of harvest. *Moniliophthora roreri* infected fruits which have not reached the spore producing stage are in many cases indistinguishable from witches' broom infected fruits.

The two diseases presently occur together in virtually all of the cocoa growing areas of Ecuador, with their individual importance changing from zone to zone (Aragundi, 1974), and also between farms within the same area, if the planting material differs in relative susceptibility. It is not clear to what extent witches' broom losses might increase if *M. roreri* were to be controlled, perhaps by sanitation. In addition to causing pod loss, witches' broom also attacks young plants and chupons, thereby interfering with renovation and regeneration of old farms. Damage to the early growth delays or prevents the formation of short, easily managed trees.

Crinipellis perniciosa is indigenous to the Amazon basin of Ecuador and presumably has long been present in the wild cocoa of the Oriente region to the east of the Andes. To the west on the coastal plain, many years of essentially disease free, successful cocoa production were abruptly terminated by the rapid spread of *M. roreri* in 1914, and *C. perniciosa* a few years later (Rorer, 1926). Vast haciendas were rendered uneconomic within a decade and many plantings were abandoned. Nevetheless, certain farmers persevered with

Disease Management in Cocoa: Comparative epidemiology of witches' broom. Edited by S.A. Rudgard, A.C. Maddison and T. Andebrhan. Published in 1993 by Chapman & Hall, London. ISBN 0 412 58190 6

cocoa and attempted to select more resistant material, with notable success in respect of witches' broom, according to Pound (1938).

Today, the only areas without appreciable losses to witches' broom are those where irrigation is needed to supplement low rainfall, for example in the El Oro and Manabí Provinces. In the main cocoa growing areas in the Los Rios and Guayas provinces, there is a long dry season, but not normally so severe as to make irrigation essential. Here, much of the main crop is lost during the 6 months wet season and many farmers rely heavily on the escape crop with peak harvests from December to March. This is true in the Quevedo zone where the two sites were located. In the Amazon region, there is only a short break in the rains and losses in cultivated cocoa are heavy.

Witches' Broom Research in Ecuador

Research on cocoa in Ecuador started in the mid–1940s with the creation of the Tropical Experimental Station at Pichilingue on the River Quevedo. A cocoa germplasm collection was established, and much of the work reported concerns the selection, breeding and evaluation of cocoa, with witches' broom tolerance high on the list of priorities (see Annual Reports: 1942–63, "Servicio Cooperativo Interamericano de Agricultura", Quito, Ecuador; 1964–present, "Instituto Nacional de Investigaciones Agropecuárias, Quito, Ecuador). The results from the trials of the Phytopathology Department are of particular interest, and it is hoped that more will be published in the near future.

Certain agronomic studies have also assessed effects on witches' broom, for example those concerning spacing (INIAP, 1972), various types of pruning (INIAP, 1974*a*), minor nutrients (INIAP, 1974*b*) and shade reduction (Maddison, 1985) but clear, consistent effects were not reported.

There is little published work on the field biology and epidemiology of *C. perniciosa* in Ecuador. Solorzano (1977), as reported in Chapter 3, used a Hirst trap to study diurnal fluctuations of basidiospores in the atmosphere, and cocoa seedlings to monitor the spread of infection from an artificial source. Seedlings exposed more than 800 m from the nearest infected cocoa at the beginning of the wet season provided strong circumstantial evidence for the movement of viable spores in large numbers over such a distance, because brooms were initiated in the plants several weeks before basidiocarps formed on the adjacent source (Evans & Solorzano, 1982). The exposure study also confirmed the finding from artificial inoculations (Evans *et al.*, 1977) that infection of flush tissue was also possible away from buds. Maddison & Mogrovejo (1984) confirmed the debilitating effects of witches' broom infection during establishment, even under temporary plantain shade.

Solorzano (1977) found that only 42% of brooms on the ground sporulated in comparison with 78% of suspended brooms, in addition, the latter produced three times as many basidiocarps (*ca.* 11/broom). Hedger (1988), also working in Pichilingue, showed that brooms soaked in certain fungicides and suspended in the canopy in nylon bags produced many times more basidiocarps than untreated brooms. Brooms in bags in the litter did not sporulate.

Over the years, various attempts have been made in Ecuador to reduce witches' broom attack by application of chemicals. Desrosiers & Bolaños (1955) reduced sporulation in the field by spraying broad spectrum fungicides and later (INIAP, 1975) emulsions of agricultural oil were used in Pichilingue to the same end. Few of the various attempts to reduce pod losses to witches' broom, *M. roreri*, and *Phytophthora* spp. by spraying pods have been successful, though Cronshaw (1979) obtained economic yield increases when hand–pollination on farmers' cocoa was followed by spraying with various copper or organic fungicides. Hand–pollination was also advocated by Edwards (1978), but timed so as to allow fruits to mature during the dry season and thereby escape disease.

Current Practices for Disease Management of Witches' Broom

The majority of traditional cocoa in Ecuador is grown under the system of low input:low output, right from its planting. There is no attempt to restrict the plant to the first or second jorquette by formation pruning, and the only commonly practised maintenance pruning is that of removal of some of the basal chupons. The result is that typical farmers' cocoa is excessively tall, usually over 7 m, and this makes harvesting slower and more laborious because of the need to use poles with extensions - most cumbersome. Very little, if any, sanitary pruning is done with respect to broom removal. Diseased pods may be harvested if the farmer believes they have beans which can be salvaged, but otherwise the majority are left on the trees. Chemicals, including fertilisers, are not applied; weeds are kept down by machete.

In certain areas, especially where cocoa is grown with supplementary irrigation, a higher level of technology is employed, which may include the use of clonal material, and formation, maintenance and occasionally some sanitary pruning. Fungicides are not applied on a regular basis, but some fertilisers may be used. Recommendations for this high input: high output system of cultivation are available in the "Manual de cultivo del cacao" (Suárez, 1987).

MATERIALS AND METHODS

The two sites in Ecuador were chosen to represent contrasting types of cocoa within the same geographical area near Quevedo in the Los Ríos Province.

Description of the Sites in Ecuador

The Pichilingue Station site consisted of hybrid cocoa which had received formation pruning, while the Pichilingue Farm site was located in typically tall, unpruned cocoa of the local hybrid complex. The Station site was to receive sanitary pruning, but the Farm site in common with other farmers' cocoa, was subject only to harvesting, basal chupon removal, and weeding. Thus the Station site was the protocol site, while the Farm site provided information on the epidemic in traditional cocoa, where brooms were not removed. The site details are compared with those from other countries in Chapter 4.

Both sites were on nearly level ground, amidst dissected terrain 60–90 m a.s.l., with deep, mature volcanic soils. They were separated by a distance of approximately 3 km, and both had other cocoa adjacent.

Pichilingue Station Site

This was located in Plantation 7A, in a part planted in 1967 with seedlings of the interclonal hybrid EET-95 x Silecia-1, at a spacing of 4 x 4 m, with permanent shade originally provided by *Inga edulis*, but virtually non-existent by 1986 (Moreira, 1989). There were 93 trees surviving in the nominal 100 tree area. The trees had been restricted to the first or second jorquette, but nevertheless some were quite tall (4-8 m). Ground cover was provided in much of the plantation by the weed *Geophila macropada*.

Preparations were started in November 1985 with the random collection of the first broom set which, because of the lack of recent sanitary pruning, was composed of brooms of various years. Records of basidiocarp production and phenology were started in week 3, 1986, several weeks after the beginning of the wet season. Harvest data were taken from week 13, and installation of the meteorological equipment was completed in week 18. However, the required removal of infected material from all the trees in the site was not carried out before the rains began in December 1985 and so the data collected during the 1986 rainy season relate to unpruned cocoa, as in the farm site.

Infected material was removed from the trees after counting in September 1986, in November 1987 and again October 1988. The 93 site trees and the 3 rows of trees surrounding the site were cleaned, and brooms were taken from the plot and burnt. The second broom set was collected and recorded from week 42 of 1986 and the third from week 44 of 1987. These two sets were composed of brooms formed in just a single wet season, namely 1985/86 and 1986/87, respectively.

Detailed recording stopped in September 1988, but harvest records were continued for a further year in the 93 trees. The method used for assessing diseased pods was as follows: at each harvest a total was obtained for every tree, and pods with recognisable symptoms (*Phytophthora* sp., *Botryodiplodia* sp. etc, and pseudostroma or sporulation of *M. roreri*) were separated. The remainder, which had symptoms that could have been either *C. perniciosa* or *M. roreri* (as yet lacking sporulation), were split and left at the base of the tree for a week before examination for the presence of pseudostroma or sporulation of *M. roreri*. Fruits that did not present such symptoms were classified as *C. perniciosa*.

Pichilingue Farm Site

The plantation was planted in 1970 approximately, amidst timber and fruit trees, at a spacing of 3.5 x 3.5 m. In 1986, the 10 detailed recording trees ranged in height from 5-9 m (average 7.5 m), and bamboo frameworks had to be constructed around them to facilitate access to the recording units in the canopy (Macias, 1988).

Brooms were not removed from the 92 trees in the site until October 1988 when detailed recording ended. The infected material used in the sets for evaluating basidiocarp production was obtained from trees neighbouring the site, the first collection being made in May 1986, and the second in August 1987. No other removal of brooms was practised in the site or in the surrounding cocoa. Results from the two sites are compared for the years

1986/7 and 1987/8, using week 40 (beginning of October) as the start because, to a certain degree, it separates the years in terms of cropping and fruit loss to *C. perniciosa*.

RESULTS AND DISCUSSION

Climate

Rainfall

Normally little rain falls in the Quevedo region between June and November – on average only 11.6% of the annual total of 2,155 mm (data for 1965–84, Pichilingue Meteorological Station). Of the remaining months January, February, March and April are particularly wet, with an average of 300–500 mm per month (Figure 7.1), and rain falls virtually every night.

Past climate data indicate an occasional year with a particularly strong "El Niño" event, such as in 1982/83, when the rains begin early and end late, with monthly totals far exceeding the normal values. All but one of the maxima shown in Figures 7.2 and 7.3 for the period November to July were recorded in 1982/83 which had a total of 5,852 mm between October and September.

One feature of the dry season is that although actual rainfall is slight, the sky is usually cloudy and there are mornings, especially in June and July, when a light drizzle or heavy mist known locally as "garúa" is present which can cause wetting of foliage and exposed pods and brooms.

The data from the two epidemiology sites showed a close similarity to those of the official station in the years under consideration, as would be expected given their proximity.

The total rainfall at the official station from October 1985 to September 1986 was 1,941 mm, against the long term average of 2,155 mm. The year 1986/87 showed 2,550 mm; with October, January, March, April and May wetter than usual, while June was rather dry in each of the years 1986–88. The 1987/88 wet season was notable for a marked dry period in March 1988, with rainfall 300 mm down on the normal, and April was drier too (Figure 7.1), giving an annual total of only 1,672 mm.

Temperature

In Pichilingue, the highest temperatures occur in the wet season from December to May (maximum, mean and minimum of 32.6, 25.9 and 21.2°C, respectively) and the lowest in the dry season in July/August (30.0, 23.5 and 18.3°C, respectively). Of the two years under study, 1986/7 was closer to the normal; 1987/8 showed a warmer period than usual from September to November, and a more rapid fall in temperatures in June.

In the following sections, aspects which are general to both sites will be considered together, followed by those which are specific to the Station or Farm site.

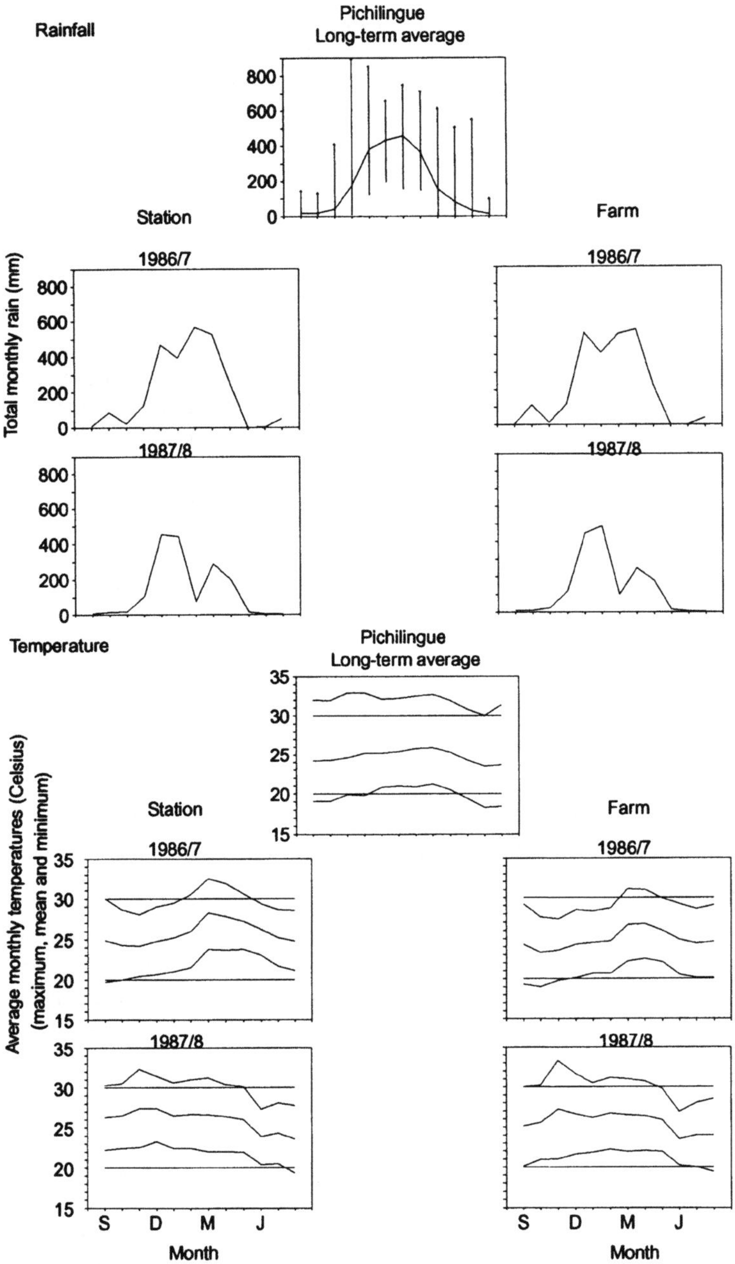

Figure 7.1. Rainfall and temperature data compared with long-term averages from the Station and Farm Sites, Ecuador.

Basidiocarp Production

General

Basidiocarp production was largely restricted to the wet season, though a few fruit bodies did appear in the 1986/87 dry season in both sites in response to a succession of isolated rains in October/November (Figures 7.2 – 7.6). A single day of heavy rain was not sufficient alone to stimulate production. In the Farm site, basidiocarps were observed occasionally during the dry season on naturally occurring material, when the sets themselves were without fruit bodies.

Although the presence of basidiocarps clearly depended on rain, the simple correlation with amount of rain per week was rather poor. This is not surprising when one considers that just a few millimetres of appropriately timed rain are sufficient to permit production in already active brooms, that too many wet days in succession may be inhibitory, and that at a given stage in the season some brooms may not yet be ready to produce, while others are already too old. Significant R^2 values of 0.52**, 0.47**, and 0.31* (*, $P < 0.05$; **, $P < 0.01$) were obtained respectively for the regressions in Set 1 Farm 1986/87, Set 1 Station 1986/87 and Set 2 Farm 1987/88. Regressions for the other set/year combinations were not significant at $P = 0.05$.

The periodicity and scale of production varied from year to year and between sites and sets, but there were a few consistent traits. For example, a given set produced less each successive year (Table 7.2), and the March/April peak was not dominant after the first year (Figures 7.2 – 7.6). In 1986/87, production was more consistent than in 1987/88. There was continuous production from week 2 until week 24 in the Station site in 1986/87 on Set 2 objects (infected material of various years in the first year of observation), while for the Farm site production on the equivalent Set 1 ran from week 51 until week 22, with only one non–productive week. By contrast, in 1987/ 88, between weeks 2 and 22, the most recent set of objects exposed showed 8 weeks without production in the Station, and 9 weeks in the Farm.

Pichilingue Station site

The relative contribution of the various materials depended on the year and set involved. For example, in the first (incomplete) year for Set 1, suspended vegetative brooms were the most productive overall with 8.1 basidiocarps per object, followed by pruned brooms on the ground and cushion brooms (Table 7.1). In their second year, 1986/87, the pruned brooms, like the other materials decomposing on the ground, produced very little, while the suspended vegetative brooms and the cushion brooms continued to produce basidiocarps, albeit at a fraction of the rate for the first year. By the third year, Set 1 showed just six active brooms (5 cushion and 1 vegetative) with a total of only 16 basidiocarps; most of the suspended tissues, and all of those on the ground, had rotted by this time. Interestingly, two of the cushion brooms had not been active before this, the third year of exposure. There was little difference in productivity between the upper and lower locations for suspended brooms in Set 1, but other sets showed greater variability.

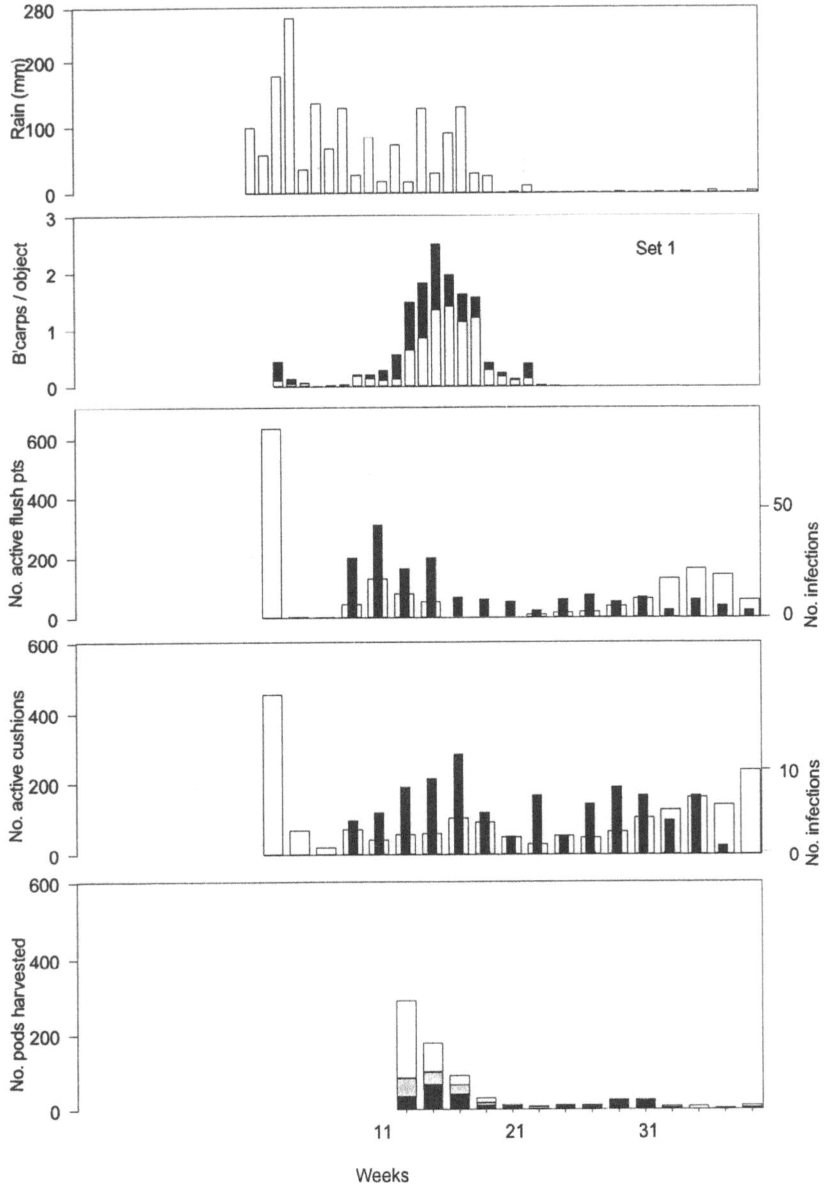

Figure 7.2. Rainfall, basidiocarp production[1], vegetative flushing and infection[2], flower cushion activity and infection[3], fruiting and fruit infection[4] in the Station Site, Ecuador during 1985/6.

[1] open columns, the most productive source type and closed columns, sum of the rest; [2] open columns, no. of healthy flushes and closed columns, no. of infections; [3] open columns, no. of active healthy cushions and closed columns, no. of infections; [4] closed columns, witches' broom infected, shaded columns, other diseases, and open columns, mature healthy fruit.

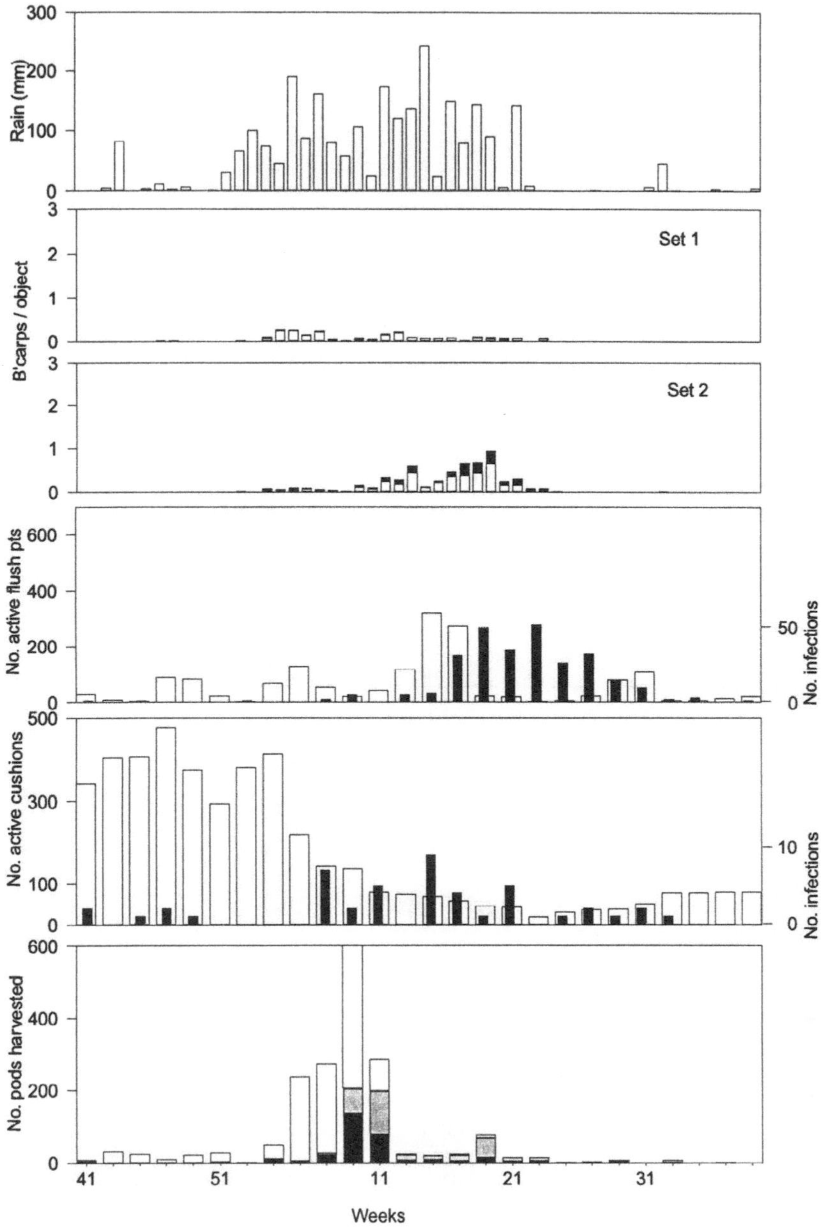

Figure 7.3. Rainfall, basidiocarp production[1], vegetative flushing and infection[2], flower cushion activity and infection[3], fruiting and fruit infection[4] in the Station Site, Ecuador during 1986/7.

[1] open columns, the most productive source type and closed columns, sum of the rest; [2] open columns, no. of healthy flushes and closed columns, no. of infections; [3] open columns, no. of active healthy cushions and closed columns, no. of infections; [4] closed columns, witches' broom infected, shaded columns, other diseases, and open columns, mature healthy fruit.

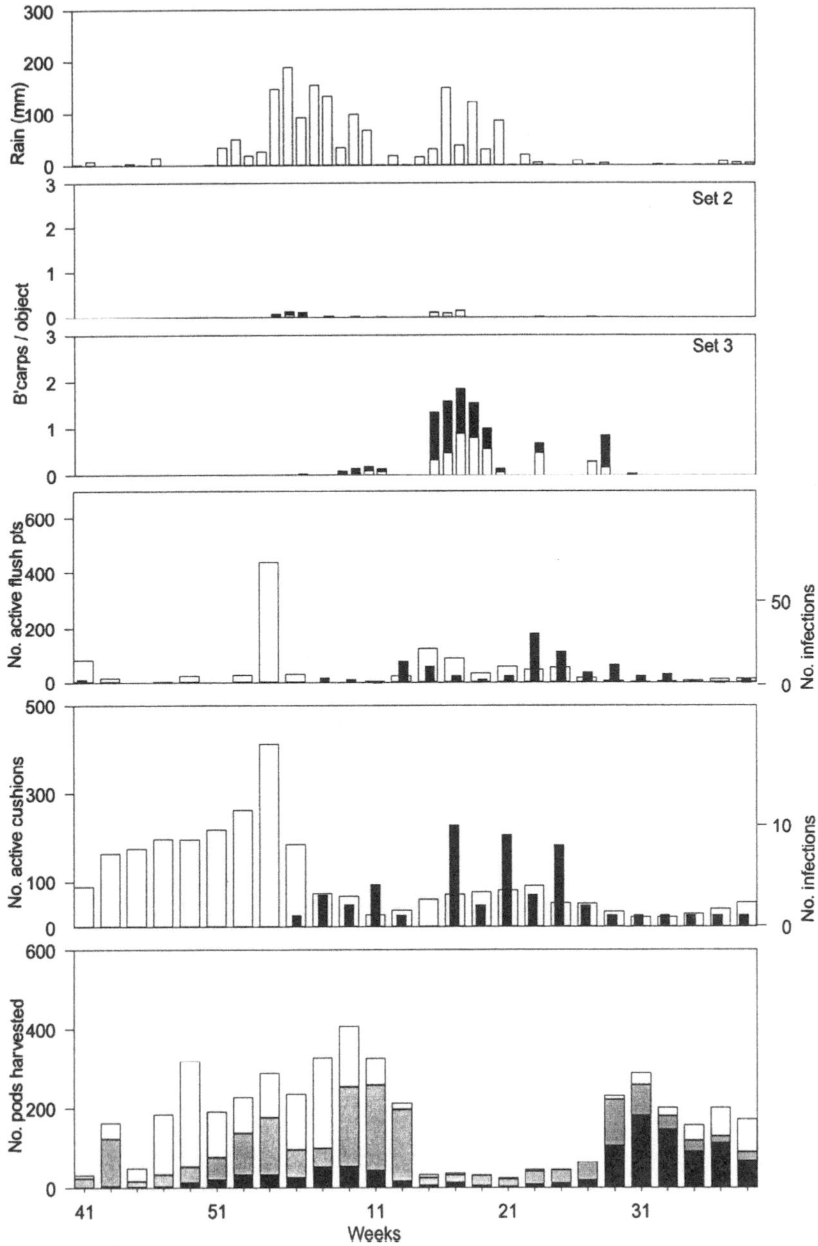

Figure 7.4. Rainfall, basidiocarp production[1], vegetative flushing and infection[2], flower cushion activity and infection[3], fruiting and fruit infection[4] in the Station Site, Ecuador during 1987/8.

[1] open columns, the most productive source type and closed columns, sum of the rest; [2] open columns, no. of healthy flushes and closed columns, no. of infections; [3] open columns, no. of active healthy cushions and closed columns, no. of infections; [4] closed columns, witches' broom infected, shaded columns, other diseases, and open columns, mature healthy fruit.

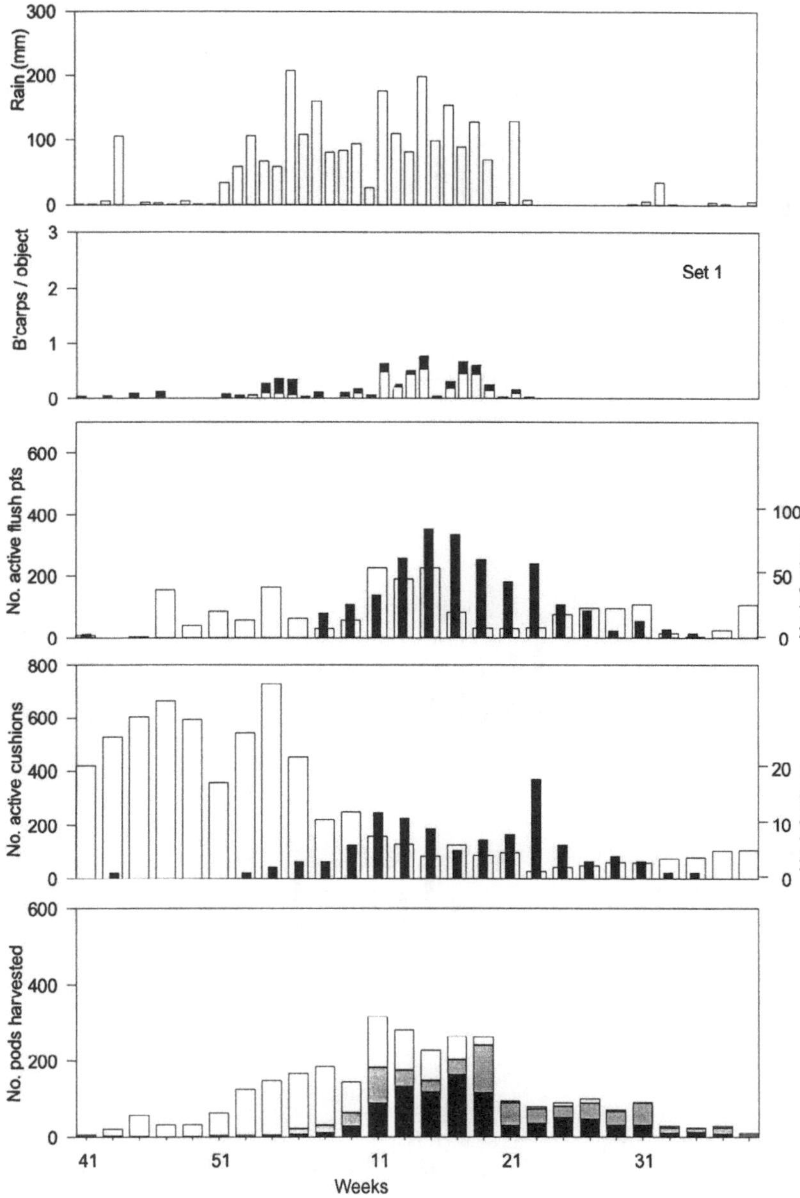

Figure 7.5. Rainfall, basidiocarp production[1], vegetative flushing and infection[2], flower cushion activity and infection[3], fruiting and fruit infection[4] in the Farm Site, Ecuador during 1986/7.

[1] open columns, the most productive source type and closed columns, sum of the rest; [2] open columns, no. of healthy flushes and closed columns, no. of infections; [3] open columns, no. of active healthy cushions and closed columns, no. of infections; [4] closed columns, witches' broom infected, shaded columns, other diseases, and open columns, mature healthy fruit.

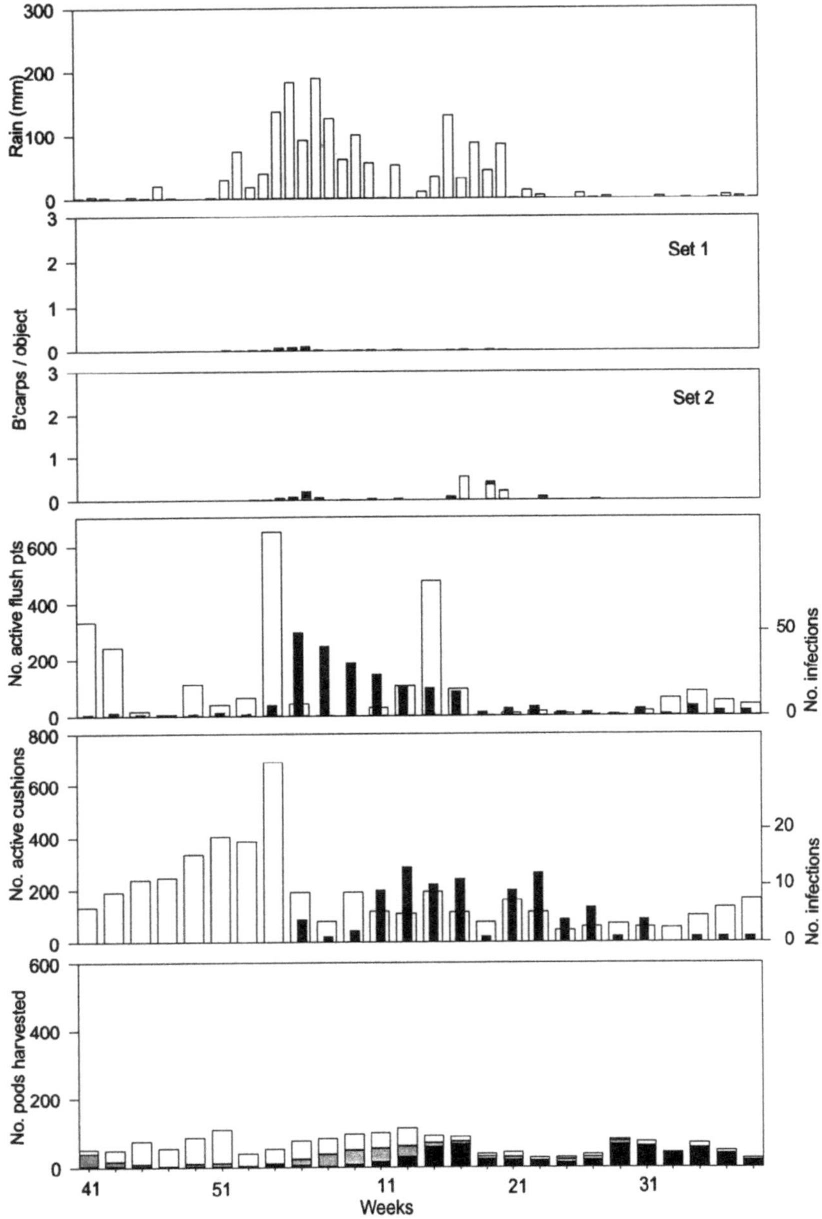

Figure 7.6. Rainfall, basidiocarp production[1], vegetative flushing and infection[2], flower cushion activity and infection[3], fruiting and fruit infection[4] in the Farm Site, Ecuador during 1987/8.

[1] open columns, the most productive source type and closed columns, sum of the rest; [2] open columns, no. of healthy flushes and closed columns, no. of infections; [3] open columns, no. of active healthy cushions and closed columns, no. of infections; [4] closed columns, witches' broom infected, shaded columns, other diseases, and open columns, mature healthy fruit.

TABLE 7.1
Totals of weekly records of basidiocarp production in various sets of diseased tissues during different years in the Pichilingue Station and Farm sites, 1986–88; totals per 100 objects

Diseased objects	Year	Suspended/on tree					On ground		
		High vege-tative	Low vege-tative	Cushion brooms	High pods	Low pods	Fallen brooms	Pruned brooms	Pods
Station									
Set 1	1985/6	723	892	138	22	76	79	327	7
	1986/7	202	173	29	2	0	2	3	0
	1987/8	1	0	8	0	0	0	0	0
Set 2	1986/7	449	279	174	6	12	4	0	4
	1987/8	29	17	43	2	0	0	0	4
Set 3	1987/8	354	513	209	0	160	23	256	5
Farm									
Set 1	1986/7	250	182	361	2	16	37	14	0
	1987/8	26	4	20	0	0	0	0	0
Set 2	1987/8	20	87	130	0	0	0	0	0

A notable feature of 1986/87 was the lower production of basidiocarps from objects on the ground relative to 1985/86 and 1987/88. This may be because rainfall was higher and more consistent in 1986/87, favouring decomposing organisms more than *C. perniciosa* and allowing better growth of the ground cover plant. Brooms in weedy quadrats rotted more quickly and produced less fruit bodies than those in open areas.

In 1987/88, Set 2 had essentially two periods of activity, weeks 2–4 and 14–16 (Figures 7.2 – 7.6), divided between cushion and vegetative brooms. Set 3 was the only set composed solely of brooms formed during the previous year, rather than during several preceeding years. Perhaps the absence of brooms which had already produced basidiocarps was responsible for the delayed start to production in Set 3, but also for the especially strong peak in April/May. Vegetative brooms had the highest percentage activity generally in the Station site, followed by cushion brooms.

Pichilingue Farm site

The Farm site in 1986/87 showed more dry season activity than the Station site, mainly from brooms on the ground and cushion brooms. In the wet season, production was numerically similar to that in the Station, coming principally from cushion brooms and vegetative brooms.

In 1987/88 the Farm site, without any obvious reason, gave few basidiocarps relative to the Station site. Cushion brooms in Set 1 in its second year and Set 2 were the most productive followed by vegetative brooms. The proportion of vegetative brooms that produced fruit bodies was only 17.5% in the Farm compared with 50% in the Station.

Flushing and Vegetative Infection

Flushing

Flushing of shoots was cyclical with four or five peaks per year (Figures 7.2 – 7.6). The two sites were more or less synchronised in a given year, although peaks occurred about 2 weeks earlier in the Farm than in the Station in 1986/87. They also differed in the number of points active and in the number of leaves per active point (Table 7.2). The Station site showed more activity in 1986/87 than in 1987/88, while the reverse was true for the Farm site. The regressions of sums of leaves on active points per tree were highly significant in both sites with R^2 values of 0.96 and 0.97, respectively. There were differences from year to year in the timing and relative height of the peaks within a site; flushing was more erratic in 1987/88 than in 1986/87. Some trees changed their degree of activity considerably from year to year while others were more consistent, for example Tree 9, which was the most active in the station site in each of the three years observed.

TABLE 7.2
Flushing and shoot infection summarised for the whole year, and for the main period with basidiocarps on the broom/pod sets in the Pichilingue Station and Farm sites, 1986–88; totals for the 50 units

Year and Site	Whole year					Inoculum period			
	Active points	No. of leaves	Leaves /active point	Total diseased points	Disease index[1]	Active points	Disease index	% active points in IP[2]	Weeks[3]
1986/7 Station	1,575	6,399	4.1	281	0.18	1,077	0.26	68	8701–8725
Farm	2,049	5,996	2.9	539	0.26	1,263	0.43	62	8651–8722
1987/8 Station	1,102	4,813	4.4	118	0.11	920	0.13	84	8801–8826
Farm	2,488	7,700	3.1	215	0.09	1,487	0.14	60	8752–8826

[1] No of diseased points/active points; [2] IP is inoculum period; [3] weeks are given as year (e.g. 87) and julian week (e.g. 03).

Vegetative infection

In both sites vegetative infection, whether expressed as totals for the 50 units or as the index "sum of infections/sum of active point counts", was greater in 1986/87 than in 1987/88. The totals/indices for the station site in the two years were considerably less than those recorded in the Farm (Table 7.2). Examination of Figures 7.2 – 7.6 reveals that in both sites two of the flush peaks coincided with the inoculum period in 1986/87, but that in 1987/88 the coincidence was less strong. Bud infections were somewhat less numerous than non-bud infections except in the Farm site in 1986/87, when the two classes were equally represented. Swellings on petioles and pulvini predominated amongst the non-bud infections, which in both years peaked in the Farm several weeks earlier than in the Station.

In an attempt to refine the analysis, the amount of flush activity during the main inoculum period was also extracted, and regressions were done of infections per tree on total active points and on active points during the inoculum period. The picture was not changed greatly by concentrating on the period with basidiocarps. None of the regressions for the farm was significant (though the R^2 values for the basidiocarp period were somewhat higher than those for the whole year) while in the station the restricted period gave a bigger R^2 value in 1986/87 (0.54* versus 0.46*), but not in 1987/88 (0.71** versus 0.83**).

The individual tree data showed that certain trees, although more active than the average in terms of flushing during the basidiocarp period, were somewhat less prone to vegetative infection. Tree 5 in the Station site had indices of 0.12 and 0.03 in the two years relative to the averages of 0.34 and 0.10. Similarly, in the Farm site, tree 4 flushed vigorously, but showed indices of 0.11 and 0.04 compared to the averages of 0.49 and 0.16. By contrast, another vigorous flusher, tree 8, was more susceptible than the average with indices of 0.52 and 0.21.

Whole-tree broom counts made in the Station site in 1986, 1987, 1988 and 1989 gave mean numbers of vegetative brooms per tree of 42.6, 77.3, 65.5 and 63.5, respectively. In the Farm site, the end-of-recording count in 1988 revealed an average of 32.0 vegetative brooms per tree (92 trees), while in 1989 the mean was 31.4.

Flowering and Cushion Infection

Flowering

As in the case of vegetative flushing, the two sites showed similar patterns of flowering through the year, but there were quantitative differences. In 1986/87, the Station site had a total count of 4,435 active cushions compared with 6,646 in the Farm, while in 1987/88 the values were 2,786 against 4,673 (Table 7.3). The differences were more marked when one considered the main periods with basidiocarps.

Flowering was most pronounced in the latter part of the dry season and in the first few weeks of the rains. Activity then continued throughout the rains at a variable but lower level (Figures 7.2 – 7.6). There was a strong relationship between the number of cushions with open flowers and the number of active cushions, with R^2 values of 0.92** and 0.91** for the regressions for the Station and the Farm respectively.

Cushion infection

Witches' broom symptoms appeared at about the same time as those on vegetative flushes, that is within 3–4 weeks of the start of the rains, and continued through the wet season and part of the dry season, until August or September (Figures 7.2 – 7.6). The few infections found in the station site in October, November and December of 1986 may have been slow developing or latent infections from the wet season or, perhaps less likely in view of their absence from the Farm site, a result of new infections from the small quantities of basidiocarps associated with the isolated rains. Although there were peaks of incidence, they seemed to have no consistent relationship with the numbers of basidiocarps produced a specific number of weeks earlier.

TABLE 7.3

Flowering and cushion infection summarised for the whole year, and for the main period with basidiocarps on the broom/pod sets in the Pichilingue Station and Farm sites, 1986–88; totals for the 50 units

Year and site	Whole year					Inoculum period		
	Active cushion	Active cushion with open flowers	Total no. diseased	% diseased in class 1[3]	Disease index[1]	Active cushion	Disease index[1]	% active cushion in IP[2]
1986/87								
Station	4,435	959	46	43	0.010	1,674	0.027	38
Farm	6,646	1,333	104	56	0.016	3,237	0.032	49
1987/88								
Station	2,786	760	51	53	0.018	1,319	0.039	47
Farm	4,673	914	90	70	0.019	2,540	0.035	54

[1] no. of diseased cushions/active cushions; [2] IP is inoculum period; [3] dead flowers and star blooms.

Infections at the station site were about half as numerous as those in the Farm, a fact which could be explained by the lower amount of cushion activity in the station. This explanation is supported by the similarity of the indices of infection derived on the basis of the total of infections in the year, divided by the total for the counts of active cushions during the main period with basidiocarps (Table 7.3).

Flower brooms (class 1) were the most common symptom, followed by chirimoyas and vegetative cushion brooms. There was considerable variation in incidence from tree to tree and from year to year, but there were certain trees which appeared to have some escape or resistance. Tree 7 in the Station site remained free from infection in two of the three years observed, partly perhaps because of its low flowering activity in these two years. Tree 9, which showed moderate flowering, had lower indices of infection than the average of the values for the ten trees, the whole year/inoculum period values being 0.004/0.009 and 0.006/ 0.015 in 1986/87 and 1987/88, respectively, compared with average values of 0.010/0.027 and 0.016/0.039.

In the Farm site, Tree 1 was very active in flowering but had lower whole year/inoculum period indices of infection than the average in the two years: 0.007/0.015 and 0.012/0.022, compared with averages of 0.022/0.032 and 0.023/0.039. Tree 4, which had revealed some resistance to vegetative infection, also showed less infection than average on the flower cushions.

The removal counts made of cushion brooms in the Station site in 1986, 1987, 1987 and 1989 gave means of 420.8, 227.2, 358.6 and 199.2, respectively, while in the Farm the values for 1988 and 1989 were 234.9 and 144.2. Thus cushion brooms always outnumbered vegetative brooms, by a factor of anything from three to ten times.

Fruiting and Fruit Infection

Fruiting

The five marked sections in the ten marked trees gave relatively little information in terms of fruiting, hence the harvest data for all the trees in a site are presented in the summaries (Figures 7.2 – 7.6). The data from the units show that setting was low from July to September and high between October and December. During the wet season there were occasional periods with moderate numbers of new pods forming. The curves for developing fruits peaked in December/January in both years, and were at their lowest in August 1987 and May/June 1988. Harvests were heaviest between November and March in the Station, and between November and April in the Farm, with a secondary peak in 1987/88 from July to September.

The Station site (93 trees) yielded a total of 1,764 fruits in 1986/87, and 4,518 fruits in 1987/88, while in the Farm site (92 trees) the respective totals were 2,985 and 1,674. Taking the two years together, the Station gave a mean of 67.5 ± 66.2 fruits per tree per year, and the Farm 55.5 ± 84.6. The respective means for the ten trees were 130.2 ± 103.4 and 97.9 ± 80.2.

The height distribution recorded for fruits in the Station was complicated by the fact that all fruits higher than 6 m were classified together, and this class dominated the distribution, which did however show a secondary peak between 3 and 4 m. All pod heights were measured in the Farm, and the distribution showed one broad peak from 4 to 7 m, with more production on the lower trunk than was the case in the Station.

Fruit infection

In the Station site, disease losses increased from 35% in 1986/87 to 62% in 1987/88, while in the Farm there was a slight decrease from 56% to 52%. The differences recorded in the number of pods available for infection probably explain to some extent the different tendencies in the two sites. In general, the percentage disease incidence recorded in the 10 trees was similar to that obtained for the whole site.

Moniliasis caused more fruit losses than witches' broom in the Station site, 18.4% versus 16.1% in 1986/87, and 38.3% versus 23.8% in 1987/88; other diseases were insignificant. In the farm site the situation was different, witches' broom dominated moniliasis with 32.0% loss versus 20.3% in 1986/87, and 32.1% versus 11.6% in 1987/88. The Farm also suffered some *Phytophthora* pod rot, namely 3.7% and 8.0% in the two years, respectively. Incidence of *Phytophthora* pod rot showed a clear decrease with height, but neither witches' broom nor moniliasis revealed obvious tendencies in this respect.

With regard to the timing of witches' broom fruit infection, the first period of removal of witches' broom pods in the Station site occurred between weeks 7 and 11 in 1986/87, and between weeks 49 and 12 in 1987/88. There was a second period of removal from week 24 until the end of recording in 1987/88, with its peak in week 30. In the Farm, the first period of removal lasted much longer than in the Station (more fruits available), and in 1987/88 the peak also came later, in weeks 14 to 16. The second period coincided more or less in the two sites.

These results, when considered along with a period of 12 weeks for the development of necrosis after initial infection, put the time of infection for the earliest necrotic fruits in the Station site at about weeks 48 and 37 in 1986/87 and 1987/88, respectively. The basidiocarp data for the Station site show that very small numbers of fruit bodies were present in weeks 46 and 47 of 1986/87, but that no basidiocarps were recorded at the time in question in 1987/88. However, in the Farm site, basidiocarps were recorded in week 34 on ground brooms within the sets.

In both sites, the main part of the first peak in witches' broom fruits came approximately 12 weeks after the minor peak in basidiocarp production seen at the beginning of January, while the July/August peak which occurred in 1987/88 was about 12 weeks after the main basidiocarp production period in April/May.

From the foregoing it is clear that although there was infection of fruits by *C. perniciosa* on occasions during the dry season, the proportion infected was low compared to that in the wet season. This gives rise to the possibility of escape from disease for trees which set pods in the dry season, especially when the pods are already three months old by the time the rains begin. In order to look for such "escapers" summaries were made of the annual production of each tree in the sites from week 40 in one year until week 39 in the next, and of the proportion of the harvest which fell in the main escape period, which was defined as week 40 to week 13 of the following year.

The results of this analysis confirmed the usual situation found with hybrid cocoa of a small proportion of the trees in the site being responsible for a large share of the production. In the Station site, with the two years' data combined, 80% of the total of fruits were produced by just 40 of the 92 trees, and when considering healthy ripe rather than total pods the number was only 35 trees. In the Farm site, with its mixture of hybrids, the distribution of the population was even more highly skewed, and the equivalent values were 24 and 20 trees from the 93 trees present.

Trees at the Station site produced 91% of their total pod harvest in the escape period of 1986/87, compared with 54% in the Farm. In 1987/88 the sites were similar, with 66% and 59% in the escape period, respectively. At the individual tree level, and considering only trees with reasonably high productivity, there were very few which produced the majority of their pods in the dry season, in both of the years under observation.

In the Station site, although many of the better producers escaped in 1986/87, most succumbed in 1987/88. Only one tree of the top twenty producers from 1986/87 showed less than 20% witches' broom infection on fruits (calculated as "WB/WB + ripe" to reduce difficulties of interpretation caused by moniliasis) in 1987/88. This tree, tree 10 of the detailed recording set, produced 90% or more of its fruits in the escape periods, and had just 5% and 19% loss to witches' broom. Unfortunately, it was rather susceptible to *M. roreri*, and lost 40% of its pods overall in 1987/88; nevertheless, it still produced 100 ripe fruits in 1986/87 and 154 in 1987/88. Several other trees showed some escape from witches' broom, accompanied by susceptibilty to moniliasis.

In the Farm, as in the Station site, few of the better than average producers escaped witches' broom in both years. One exception was tree 20, which produced 83% and 86% of its fruits in the two escape periods, had losses of 15% and 19% to witches' broom, and 17% and 16% to moniliasis. It yielded 55 and 81 ripe pods in the two years. Tree 58 was more productive, and still showed appreciable escape (65% and 75%), moderate losses to witches' broom (32% and 21%) and moniliasis (14% and 4%), with good healthy pod production (161 and 59) in the two years 1986/87 and 1987/88.

Regarding actual resistance to fruit infection during the wet season, as distinct from escape, analyses were done of the production of healthy and diseased fruits in the period from week 15 to week 39, 1987 and week 14 to 38 in 1988. The Station site produced so few pods in the 1986/87 season that no conclusions can be drawn about possible resistance in that year. In 1987/88, there were no really outstanding trees, though the better examples had percentage witches' broom losses of 30–40% compared to the mean (for trees producing more than 10 pods in this period) of 49%. In the farm, tree 88 lost only 25% relative to the mean of 50% in 1986/87, and 36% against 69% in 1987/88, with ripe pod production in the two years of 25 and 18, respectively.

CONCLUSIONS AND DISEASE MANAGEMENT IMPLICATIONS

1. The "6 months wet, 6 months dry" description of the climate in the Quevedo zone is useful when examining the epidemic, because much of the development is initiated during the wet season and then limited by the dry season.

2. However, isolated rains in the dry season can be important by giving conditions suitable for basidiocarp formation and infection. Resultant fruit losses can be appreciable as a considerable proportion of the crop is formed in the dry season.

3. Just two days of rain in the dry season can be enough to permit basidiocarp formation on brooms that have produced previously. Such brooms are also responsible for the early wet season peak of basidiocarp production which, although small, may in some years cause much infection in shoots, cushions and fruits because of the availability of expanding tissue at that time.

4. The main peak of basidiocarp production is centered on April, and its effects vary considerably, according to the phenology activity of the trees at this time.

5. Most brooms produce basidiocarps during the first two years; very few continue into the third year. Brooms on the ground, especially pruned brooms, are an important source of basidiocarps in certain years, though they rot sooner than those on the tree, particularly if

there is reasonable weed cover. Suspended vegetative brooms are the most prolific in the long term, but their contribution to the total population of basidiocarps depends also on their numbers in the site. Cushion brooms outnumber vegetative brooms by between 3 and 10:1, and because of this they may have similar, or greater importance overall as basidiocarp producers.

6. Variations within the basic pattern of phenology lead to considerable quantitative differences from one year to another, in flushing, flowering and fruiting. The two sites, although showing the same approximate periodicity in certain parameters, often differ widely in fruit production, and infection of the various tissues.

7. The long period during which symptoms can appear makes it difficult to relate infection to basidiocarp production, phenology and climate, but circumstantial evidence suggests that relatively few basidiocarps can cause considerable infection.

8. The appreciable variation between cocoa trees encountered means that a small proportion of the trees, generally the vigorous trees, are responsible for a disproportionate amount of the phenological activity, and infections.

9. Disease escape is important, but few trees produce most of their crop consistently in the dry season. Most of the productive trees are year–round producers that have moderate or poor indices of infection, but nevertheless yield well because of their partial escape. A few escapers of reasonable productivity are present in the populations studied, as are one or two trees with somewhat higher than average wet season resistance to witches' broom.

10. The most urgent recommendation for disease management is that the selection and breeding of more resistant material and better escapers are encouraged to the full.

11. In the shorter term, looking at the existing traditional cocoa, it has to be admitted that it is largely unsuitable for normal rehabilitation measures due to its height. Drastic pruning schemes may prove economic, if the expected increase in yield can be maintained for more than a few years.

12. Given a background of short, well managed trees, the following recommendations for management could be made, though the necessary proof of their usefulness is still lacking.

12.1 If light structural pruning is carried out in July/August with the aim of stimulating flowering and therefore disease escape, it should include removal of as many brooms as possible, but especially the more easily removed old brooms. Otherwise broom removal could be left until September, to be followed in both cases by a further removal in October/November.

12.2 Brooms should not be left in the plantation.

12.3 Low weed growth in the plantation is desirable.

Chapter 8

COMPARATIVE EPIDEMIOLOGY STUDY: GRENADA

C. Williams & F. James

INTRODUCTION

Importance of Cocoa and Witches' Broom in Grenada

The first serious attempts to cultivate cocoa in Grenada began around 1714, so the crop must have been introduced earlier (Cruikshank, 1970). A census taken in 1700 found that there were 150,000 cocoa trees planted on the island. The first record of cocoa being exported from Grenada was in 1763, when there were nearly 300 ha of land under cocoa cultivation. The dominant crop, sugar, occupied about 13,000 ha at that time.

Between the late 1700s and mid 1800s, the old plantation system gradually disintegrated as the number of sugar and coffee estates declined, and cocoa estates became more important. Cocoa's real expansion into the major estate and export crop began with Emancipation in 1838, with the change in labour status. By 1860, cocoa had replaced sugar, cotton and coffee as the major agricultural industry in Grenada.

Production reached a maximum of 6,300 t in 1914. There followed a gradual decline, with the average from the last five years being about 1,500 t. The decrease was attributable to many causes, amongst them a hurricane in 1921, the Great Depression, conversion to other crops, and the age of the cocoa trees. The worsening conditions forced farmers to petition the colonial Government in 1938 urging for assistance to maintain the industry, as a result of which the Grenada Cocoa Rehabilitation Scheme was established in 1940.

In the following three years, a local programme of selection generated 79 clones (G.S. : Grenada Selection) which were propagated, tested and distributed. The G.S. clones formed the basis of the rehabilitation of the industry, with distribution of rooted cuttings starting in 1946. A further hurricane in 1955 set the planting programme back, but the increasing demand for material led to the preferred rooted cuttings being supplemented with budded seedlings. The main replanting scheme lasted until 1962, and the majority of fields on the island are still of clonal trees.

Witches' broom was first reported in Grenada in the late 1930s, although there was no specific description of where and when it appeared. In 1948, the Department of

Disease Management in Cocoa: Comparative epidemiology of witches' broom. Edited by S.A. Rudgard, A.C. Maddison and T. Andebrhan. Published in 1993 by Chapman & Hall, London. ISBN 0 412 58190 6

Agriculture commissioned an islandwide survey of the disease, involving field inspections and identification of infected trees. Marked trees were supposed to be cut down and burnt by the farmers. No records have been found of pod losses or amounts of brooms from old trial or farm data. The only disease records to be found in the old trials are for *Phytophthora* pod rot. Present pod loss due to witches' broom disease has been estimated at between 5 and 10% for the whole island. It is possible that losses were never significantly greater, so the disease was not recorded in trials or mentioned as a contributory factor in yield decline.

METHODS

Description of the Grenada Site

The site was located in the central eastern part of the island in a mature field of a commercial farm close to Boulogne, where the trees were approximately 30 years old. The trees were clonal types of ICS origin established as rooted cuttings, although the identity of individual plants was not known. The area was originally planted with ICS material as an observation trial. The site was close to the propagating station of the Grenada Cocoa Association where total rainfall is recorded, and the nearest official meteorological station was 3 km away at the Mirabeau Agricultural Station. The trees used in the experiment formed a continuous block with sparse overhead shade. The field had been regularly harvested and normal practices such as weed control and chupon removal carried out before the IWBP started. The farmer had also been regularly removing the few brooms that formed prior to the experiment.

Site Installation

The site was prepared in July 1987 in accordance with the IWBP specifications by marking trees and selecting the 10 principal sample plants. Vegetative brooms were removed from all trees, together with all mummified diseased pods. The required number of vegetative brooms for the various sets were then collected by supplementing removed site-brooms with others from outside the site. High *Phytophthora* pod rot and low witches' broom incidence meant that it was impossible to select enough mummified pods known to have been infected by *Crinipellis perniciosa* for the sets for monitoring basidiocarp production. The required number of infected cushions on the 10 sample trees were marked, and then all other cushion infections were removed. Any surplus infected material was removed from the site.

The IWBP meteorological equipment was placed 50 m away at the edge of the field and records were available from the middle of the 1987 rainy season (week 33) through to late in the 1989 rainy season (week 49). Thus, data were only available for one complete rainy season; 1988. Most methods for recording data at the Grenada site were laid out in the introduction to the comparative epidemiology experiment (Chapter 4). Pod harvest records on the 64 site trees were taken every 4–6 weeks rather than at strict 2 week intervals. A complete set of new brooms was placed on the ground in May 1988, but the canopy sets were left. All sets were replaced in May 1989 from within the field, and the original sets were removed from the site.

RESULTS AND DISCUSSION

Climate

Rain was generally concentrated into the period June to December (Figure 8.1), and means of 100 mm or less were recorded in the months January to May. Variations about the means were great for all months. Data from the two years of the comparative epidemiology experiment show a similar overall pattern of wet and dry periods. The highest monthly totals were about 350 mm in both years, recorded in September (1987) and October (1988). Months with significant quantities of rain did occur in the drier period, such as March 1989 (128 mm). The least monthly rainfall of 10.7 mm occurred in May 1988. The total rainfall (September–August) was 1,671 mm in 1987/8 and 1,794 mm in 1988/9.

Long–term average temperature data showed little seasonal variation for any of the three variables, minimum, maximum and mean (Figure 8.1). All three were slightly lower between November and May, but the average minimum never fell below 20°C and the maximum did not pass 30°C at any time. Temperatures during the experimental period were similar to the long–term data with the exception of August and September 1988 when temperatures were lower than expected.

Basidiocarp Production

Suspended vegetative brooms were the most prolific sources of basidiocarps throughout the experiment. A small amount of basidiocarp production occurred on diseased cushions, and even less on vegetative brooms on the ground. Neither of the latter two sources produced basidiocarps in a week when no basidiocarps formed on suspended vegetative brooms.

Suspended vegetative brooms

Observations started in week 32 of 1987, and although the first basidiocarp was seen in week 37, production was not consistent until week 43 (Figure 8.2). There were two principal periods when less basidiocarps were produced, the first from weeks 14–27 of 1988, and the second from weeks 4–10 of 1989. Basidiocarps were observed on suspended brooms in 92 out of the 120 weeks for which data were available, on 65% of the sample of 200 brooms in 1987, 99% in 1988 and 95% in 1989. A highly significant correlation for numbers of basidiocarps was obtained between suspended brooms in the upper and mid–canopy in both 1987/8 ($r = 0.88$) and 1988/9 ($r = 0.77$). A total of 4,488 basidiocarps formed on the 200 brooms during the two years of the experiment. The average basidiocarp index was 22.4, 14.9 of which formed on the first sets up to early 1989, and 7.5 of which formed on the second sets in the rest of 1989.

Cushion brooms

Basidiocarps were recorded on diseased cushions in 23 out of the 120 weeks of the experiment. Production was largely confined to weeks 37–11 of 1987/8 and weeks 11–48 of 1989. Very few basidiocarps formed during 1988. The total number of basidiocarps found on the 200 cushions was 166, 3, and 38 for the three rainy seasons, on 19, 2 and 28 cushions, respectively. Each diseased cushion produced a mean of 7.4 basidiocarps over the whole

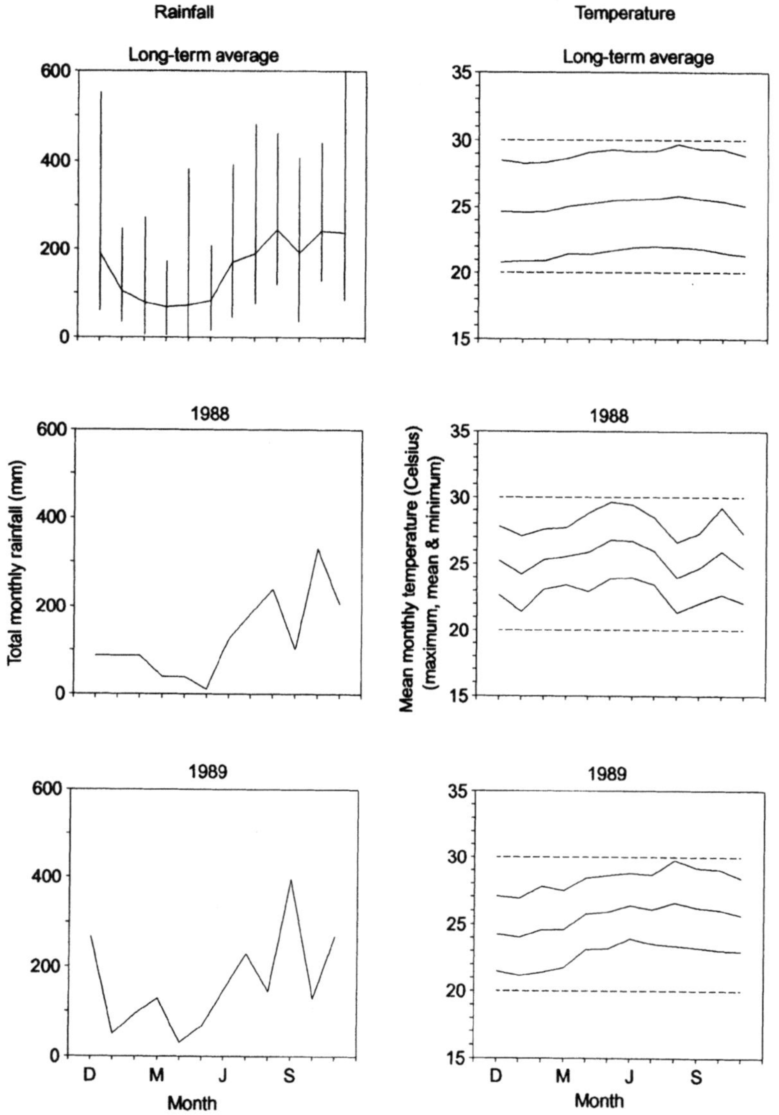

Figure 8.1. Rainfall and temperature data compared with long–term averages from the Grenada site.

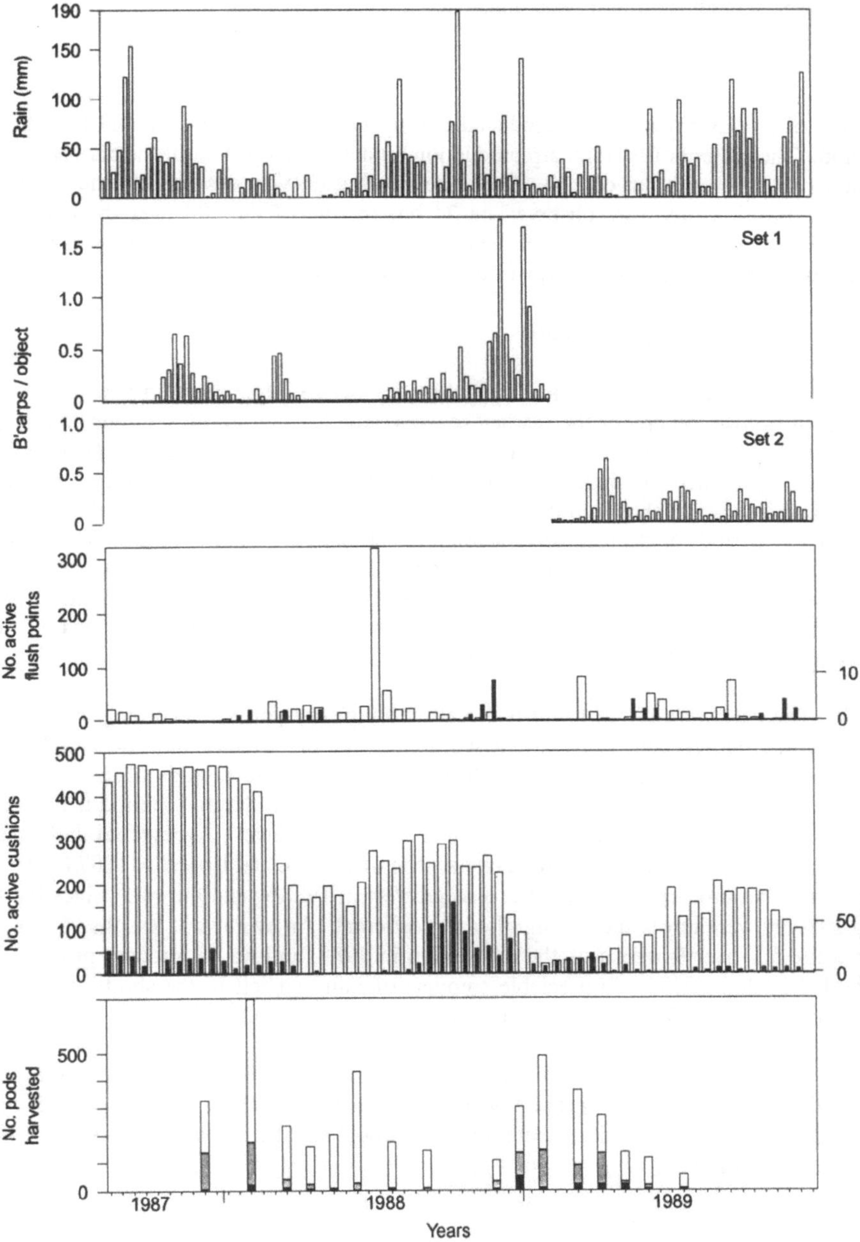

Figure 8.2. Rainfall, basidiocarp production, vegetative flushing and infection[1], flower cushion activity and infection[2], fruiting and fruit infection[3] in the Grenada Site in 1987–1989.

[1] open columns, no. of healthy flushes and closed columns, no. of infections; [2] open columns, no. of active healthy cushions and closed columns, no. of infections; [3] closed columns, witches' broom infected, shaded columns, other diseases, and open columns, mature healthy fruit.

experiment, though the figure was higher for the 1987/8 season.

Vegetative brooms on the ground

A total of 84 basidiocarps was produced during the experiment on the 100 brooms on the ground. The pattern of basidiocarp production broadly matched that of the diseased cushions with 20 active weeks out of 120. Only one basidiocarp was observed in the 1988/9 rainy season. Basidiocarps were observed on 24 brooms out of the 100 over the two years, of which 8 produced 41 basidiocarps in 1987/8 and 20 produced 32 basidiocarps in 1989. Only four brooms produced basidiocarps in the first year did so in the last.

Climate and basidiocarp production

In only one week of the first ten after the start of the experiment in week 32 1987 did basidiocarps form on any of the marked sources, and 597 mm of rain fell during that period. After week 43 1987, numbers of basidiocarps formed on the suspended vegetative brooms were broadly related to amounts of rain. The break in production early in 1988 reported above coincided with three months of less than 30 mm, when the mean weekly rain was 13 mm. Three periods of reduced basidiocarp production in 1989 occurred at times of decreased rainfall in January/February, April, and August. Although only 35 basidiocarps formed on 200 suspended brooms between weeks 4 and 10 of 1989, there was no clear break in production as occurred in 1988. The driest week in that period had 4 mm of rain, and the average over the seven weeks was 17 mm.

The data were divided into five main rainy or dry periods according to quantities of basidiocarps on suspended brooms, based around the dry periods in 1988 and 1989. The rainy seasons of 1987 and 1988 were further subdivided into periods of sporulation or no sporulation (Table 8.1). The longer dry period in 1988 was the only period that had higher temperatures and lower relative humidity than the overall mean. Average rainfall during the recording period was greatest in the rainy season of 1988/9, and lowest in 1987/8. The mean number of basidiocarps recorded was also greatest in 1988/9, and the two values for 1987/8 and 1989/90 were lower and not significantly different. An average of less than 5 mm of rain per week fell over 9 weeks of the dry season in 1988, and no basidiocarp production was recorded in that period. Appreciable amounts of rain that fell in the shorter dry period of 1989 allowed fructification to continue during the drier period.

Relationships between rain and basidiocarp production on cushion infections or vegetative brooms on the ground were less clear than those for suspended brooms. Basidiocarps were observed on these sources throughout the inoculum periods in 1987 and 1989, but not in 1988.

Flushing and vegetative infection

Flushing

Shoot activity was minimal in late 1987, after a small amount of growth immediately at the start of recording in August 1987 (Figure 8.2). Timing of growth was similar in 1988 and 1989, with three identifiable active periods or flushes in March, June/July and

September/October. The June flush of 1988 was the most intense of the experiment, with 319 active points registered at one observation. It occurred at the beginning of the rainy season, and the absence of such a large flush in 1989 was probably caused by unusually high rainfall at that time. The other peaks of activity were 37 and 15 points for March and October of 1988 respectively, and 82, 49 and 75 points in order for the three flushes in 1989. The mean number of active points registered for each tree over the whole experiment was 109, from the 50 active points initially marked (spread between five units). Tree 6 had only 13 active flush points in the two years, and Tree 10 was the most active at 207 points.

TABLE 8.1
Mean values for various climate variables and associated mean numbers of basidiocarps at different times in the experiment, Grenada site.

Time		Duration (weeks)	Rain mm/week	Maximum temp.	Minimum temp.	Minimum r.h.	No. b'carps
Rainy season							
1987/8	no sporulation	11	54.3	28.9	23.4	68.1	0
	sporulation	25	27.0	27.6	22.2	66.6	19.5
1988/9	no sporulation	7	36.9	28.9	23.8	66.8	0
	sporulation	28	46.2	27.3	22.0	71.5	35.6
1989[1]	sporulation	29	35.7	28.3	23.1	70.6	20.3
Dry season							
1988		9	4.7	29.4	23.5	61.6	0
1989[1]		7	19.2	27.9	21.5	69.0	2.6

[1] new broom set.

The degree of correlation of vegetative growth between trees was evaluated using the pooled totals per assessment of active points from the five units during the course of the experiment ($n = 57$). Two trees (nos. 2 and 6) showed no significant correlations with any other tree, but tree 6 showed almost no shoot activity. Six trees (nos. 1, 3, 4, 7, 9 and 10) were highly correlated with each other ($P < 0.001$). That group was moderately correlated with a second group of the two remaining trees (nos. 5 and 8).

Vegetative infection

A total of 31 new infections was observed during the experiment in the 50 vegetative units spread through the sample trees. All but one of these were terminal brooms or axillary brooms without any associated stem or leaf base swelling. The brooms formed either in weeks 40–46 of 1988 and 1989, or in a defined period earlier in each year; weeks 4–16, 1988 and 18–22, 1989. Four trees (nos. 2, 8, 9, and 10) had no infections in the marked units during the experiment. Twenty-five of the brooms (80%) formed on trees 1, 4 and 5. The number of vegetative brooms on the trees was not related to the amount of shoot activity.

The peak flush of shoot growth in June 1988 occurred during a period without basidiocarp formation on any marked source. Brooms formed in the units were mostly derived from infections in a shoot flush 5–10 weeks previously. A disease index was calculated as "sum of infections/sum of active point counts" for each sporulation period, and pooled data from the ten trees indicated that shoot infection increased marginally from 0.05 in 1987/8 to 0.07 in 1988/9. Data were not available for the whole 1987/8 period of basidiocarp production, and it was not possible to draw firm conclusions on disease progress by comparing the indices. Certainly, the index was very low in both cases, and did not increase greatly after two seasons without sanitation.

Flowering and Cushion Infection

Flowering

A clear annual pattern was found in cushion activity, related to the annual climate cycle (Figure 8.2). The number of active cushions decreased from January to a minimum in March/May, before increasing to a maximum around October (week 40) in all three years. The period of least cushion activity was slightly earlier in 1989 than 1988 for no clear reason. The maximum number of active cushions observed in each year decreased between 1987 and 1989 for all ten trees, and for the pooled totals (Figure 8.2). The number of active cushions within the units varied greatly between the trees.

Numbers of cushions with open flowers were highly correlated with numbers of active cushions. Synchronization of cushion activity between trees was evaluated by correlation analysis, and the coefficients for all possible combinations were significant at $P < 0.001$.

Cushion infection

Infections on cushions were placed into the different classes as described in the experimental protocol. The majority of new infections (65%) recorded during the experiment occurred as diseased flowers (Class 1), with an average of 49 per tree. Twenty–one diseased cushions with vegetative brooms were also recorded, and overall a mean of 75 diseased cushions were found on each tree. The average maximum number of healthy active cushions per tree was 48 per tree in 1987, 35 in 1988, and 23 in 1989. The decrease in healthy cushion activity might have been caused by cushion loss due to necrosis after infection and broom formation.

Most diseased cushions appeared in the period between September (week 35) and May (week 20), from cushions that became infected during the peak of flowering and the most intense period of basidiocarp production (Figure 8.2). Incidence of disease in September–November 1989 was greatly reduced from the same time in 1988, which may have been due to the smaller numbers of healthy active cushions.

The number of diseased cushions per tree was related to the number of healthy active cushions, giving a linear regression with data from ten trees that was highly significant (F ratio = 59, $P < 0.01$). The more actively flowering trees had by definition more susceptible meristems, and they also had more diseased cushions.

Fruiting and Fruit Infection

Fruit set was recorded in the marked branch units, and was found to follow a clear annual pattern. Most new cherelles appeared between weeks 30 and 48 (August and November) in all three years of the experiment, when flower cushion activity was most intense. Total set in that period declined from 1987 to 1989, with 71, 62 and 28 pods recorded in the three years. The reduction in fruit load might have been related to the decrease in healthy cushions reported above. Healthy pods would be expected to mature at least 20 weeks after fruit set, and harvests from the site trees were indeed greatest in the period December to May.

Pods diseased with witches' broom accounted for less than 5% of all pods harvested in the site, and almost 20% of harvested fruit were infected with *Phytophthora* pod rot. Witches' broom diseased fruit were recorded during most of the two years of the experiment, but were concentrated into the period December to May of 1987/8 and 1988/9. The timing of fruit set meant that most developing pods would be potentially exposed to inoculum throughout their period of susceptibility, even in years with a clear dry season.

Basidiocarp production on the marked sources was most intense during and after the main fruit set, and climatic conditions appeared to have been favourable for spore release and infection. It is difficult to account for the low pod disease incidence, unless the low density of basidiocarp sources in general, apart from the marked sets, was critical. Few brooms were present on trees in the site area either before or after the experiment.

CONCLUSIONS

1. There was a more even distribution of rainfall in the 1989 dry season than in 1988, which could have accounted for the greater basidiocarp production in 1989.

2. Brooms on the ground hardly produced any basidiocarps due to their rapid decomposition.

3. Most vegetative shoot growth did not occur during the sporulation period, so that the volume of available tissue was small. The few vegetative brooms that formed did so following periods of intense basidiocarp production and high rainfall.

4. The most intense flower cushion activity occurred during the main period of basidiocarp production. The maximum number of active cushions recorded decreased through the three years of the experiment. The numbers of infections of the different types were high, and might have been associated with the decline in overall cushion activity. Total pod production was lower in 1989 than in 1988.

5. Most pod set occurred during the peak of basidiocarp production, and the young pods were exposed to possible infection periods throughout their susceptible stage. There was no clear explanation for the low pod losses recorded during the experiment.

MANAGEMENT IMPLICATIONS

The principal threat to cocoa production in Grenada in sites such as the IWBP experiment is indirectly through accumulated cushion infection. There was no direct evidence that reduced cushion activity decreased pod set, but the risk should be avoided. Certainly the rate of pod loss alone would not justify any action being taken to control the disease.

In areas of the island where a programme of phytosanitation is needed to control cushion disease, all brooms should be removed before the onset of the rainy season in April/May. The brooms could be left in the fields as long as they are not gathered into heaps, because decomposition is so rapid. If brooms are heaped then they should be covered, otherwise brooms should be taken out and burnt.

No chemical would be necessary specifically against witches' broom, but spraying is recommended against *Phytophthora* pod rot. The treatment consists of a single massive prophylactic spray of copper in early October.

Chapter 9

COMPARATIVE EPIDEMIOLOGY STUDY: TRINIDAD

S. Mohan & T.N. Sreenivasan

INTRODUCTION

Importance of Cocoa and Witches' Broom in Trinidad and Tobago

The cocoa industry of Trinidad and Tobago has been in steady decline since the years of peak production up to 1920, when it was the fourth largest producer world-wide (Wood, 1985). The appearance of witches' broom disease reduced production and added to labour costs at a time when the world cocoa price was low, and some less profitable estates were already experiencing difficulties. Then, oil and petroleum products started to dominate exports, and wages for farm workers had to rise to keep pace with the industrial sector. A gradual abandonment of cocoa plantations started in Trinidad in the 1930s in the areas of lower soil fertility, and accelerated with the labour crisis until only the most favourably situated plantations survived. Annual cocoa production from the islands peaked at around 35,000 t and has fallen to about 3,000 t currently exported. The chances of reversing this decline are slim according to Wood (1985), and the cocoa industry may remain at its present level. The Trinidad & Tobago Cocoa Industry Board has embarked on a rehabilitation programme aimed at improving cocoa management and replacing old trees with a standard set of TSH (Trinidad Selected Hybrids) clones.

The dominance of witches' broom stimulated an intensive research programme that has covered all aspects of cocoa agronomy, breeding, pest management and economics. Much of the early knowledge of witches' broom came from such work, as is shown in Chapter 2, and a comprehensive review of relevant literature from Trinidad and Tobago would be repetitive. However, a few points relevant to this specific chapter will be mentioned.

Witches' broom disease was first reported from Trinidad in May 1928 (Stell, 1928). When the disease was discovered in eastern Trinidad, it had apparently been present for a number of years and was sufficiently established that eradication was impractical (Baker, 1953). It was initially confined to two areas in the north east of the island with high annual rainfall (Freeman, 1928), but the disease spread rapidly through the island. Stell (1937) reported that losses on a government estate increased from 1% in 1932 to 2% in 1934, 6% in 1936 and 33% on neighbouring plantations. By the late 1930s it was apparent that losses due to the disease were becoming serious over a wide area, to the extent that production in

Disease Management in Cocoa: Comparative epidemiology of witches' broom. Edited by S.A. Rudgard, A.C. Maddison and T. Andebrhan. Published in 1993 by Chapman & Hall, London.
ISBN 0 412 58190 6

1936 was half that of the peak years 1921 to 1929 when it was between 22,000 and 35,000 t (Baker & Holliday, 1957). Baker *et al.* (1941) reported that high proportions of pods were lost where the disease became severe, and losses of 40 to 45% of all pods harvested in a year were common. Disease incidence in individual harvests was sometimes as high as 70 or 80%. In a plot at River Estate, records showed that during 1934–1940 the average number of brooms per tree was 51. This number increased threefold by 1941, and in 1942–1943, 336 brooms per tree were recorded (Baker & Holliday, 1957). Pod infection rose from 10 to 68% from 1942 to 1944. Disease intensity at this site had reached its maximum by the end of 1944, and it was probably the worst affected area in the island (Baker & Dale, 1944). A similar epidemic stage was reached in parts of western Tobago in 1951. Baker *et al.* (1941) observed that basidiocarps were formed throughout the year in wetter parts of Trinidad, but that formation was reduced or inhibited for up to 6 months of the year in drier areas. Maximum disease development was associated with moderate rainfall (1,500–2,300 mm), and disease incidence was reduced in the high rainfall areas (more than 3,000 mm) found in the Northern Range (Baker & Crowdy, 1943).

Recent assessments of the disease in Trinidad by Laker *et al.* (1988*a*) found intensity to be lower than the epidemic levels reported above. This was mostly attributed to the introduction of the resistant TSH in the replanting programme started in the late 1950s. Other varieties in germplasm collections were often as heavily infected as previous authors had recorded.

In Trinidad, the recognised "flush" periods of vegetative growth occurred in the dry season during February or March and again in September or October, with smaller flushes at other intervals (Baker *et al.*, 1941). Pyke (1933) and Humphries (1942) reported five flushes per year on old trees, with flushing intervals of 60 to 90 days, and a close relationship between flushing and maximum temperature was noted. However, 3 to 4 year old trees flushed more often, 6 to 8 times at intervals of 50 to 60 days (McKee, 1942). All the above work was conducted on trees either of the old "Trinitario" type or the Imperial College Selections (ICS) clones developed in Trinidad in the 1940s. Most commercial cocoa presently on the islands is derived from rooted cuttings of the ICS and TSH materials mentioned above. Patterns of shoot growth and flowering of the TSH trees need not be similar to those found by workers before 1945. The main harvest pattern has not changed (Hardy, 1954), and few flowers and new pods are known to form from January to March. Most flowering is known to occur during the early wet season, with pod setting in June/July so that the main crop matures from December to February.

METHODS

Description of the Trinidad Site

The site was located in a mature assessment trial of TSH materials established in 1978 as part of a replicated experiment at several sites throughout the northern half of the island. The replicate planting at Santa Cruz is in a valley in the Northern Range known from historical records to be an area of moderate to high disease incidence. It is close to a Ministry of Agriculture Station (La Pastora) where total rainfall is recorded, and the nearest official meteorological station is about 7 km away at the St Augustine campus. The trees used for recording data were in two contiguous blocks of TSH clones, 919 and 1188, planted with sparse overhead shade on the level valley floor. The trees had been regularly harvested, and normal practices such as weed control and chupon removal were carried out before the IWBP started. Brooms had been removed every year in April or May up to and including 1985.

Site Installation

The site was prepared in April 1986 by marking trees and erecting a fence around the plots. Vegetative brooms were removed from all trees and counted, before placing the various broom sets in position in the canopy and on the ground. The required number of infected cushions on the 10 sample trees were marked, and then all other cushion infections were removed from the trees in the plot for counting. Any surplus diseased material was taken away from the site. Diseased pods were placed in position for basidiocarp counts as they were harvested, and the set was complete in week 40. The other plots in the trial were also cleared of brooms at that time. The Santa Cruz river flooded the valley and inundated the IWBP site in June 1987, and all basidiocarp sources on the ground were lost. They were replaced by material collected within the site. The IWBP meteorological equipment was placed 65 m away on the edge of the TSH trial area, and records are available from June 1986 to May 1988. All methods for recording data at the Santa Cruz site were as laid out in the introduction to the comparative epidemiology study (Chapter 4, Part 1).

RESULTS AND DISCUSSION

Climate

Rain was concentrated into the period May to December in the long–term data (Figure 9.1), and means of less than 50 mm of rain were recorded in the months January to April. Data for the two years of the comparative epidemiology experiment show a similar overall pattern of wet and dry periods. The highest monthly total was recorded in November in both years, and July was also particularly wet. August and September tended to be slightly drier, especially in 1987. The total rainfall (May–April) was 1,465 mm in 1986/7 and 1,346 mm in 1987/8.

Long–term temperature averages showed little seasonal variation for any of the three variables, minimum, maximum and mean (Figure 9.1). All three decreased slightly during the dry season, but the minimum never fell below 20°C and the monthly maximum did not rise above 32°C throughout the year. Minimum temperatures during the experimental period were slightly lower than the long–term normals, especially from January to April. Mean temperatures were similar to the normals, and monthly maxima differed only in February and March when they were higher.

Basidiocarp Production

Suspended vegetative brooms were the most prolific sources of basidiocarps in both years of recording, and diseased cushions were the only other material on which significant fructification occurred. A very small number of basidiocarps were observed on suspended diseased pods. Periodic measurements of randomly sampled basidiocarps showed that the

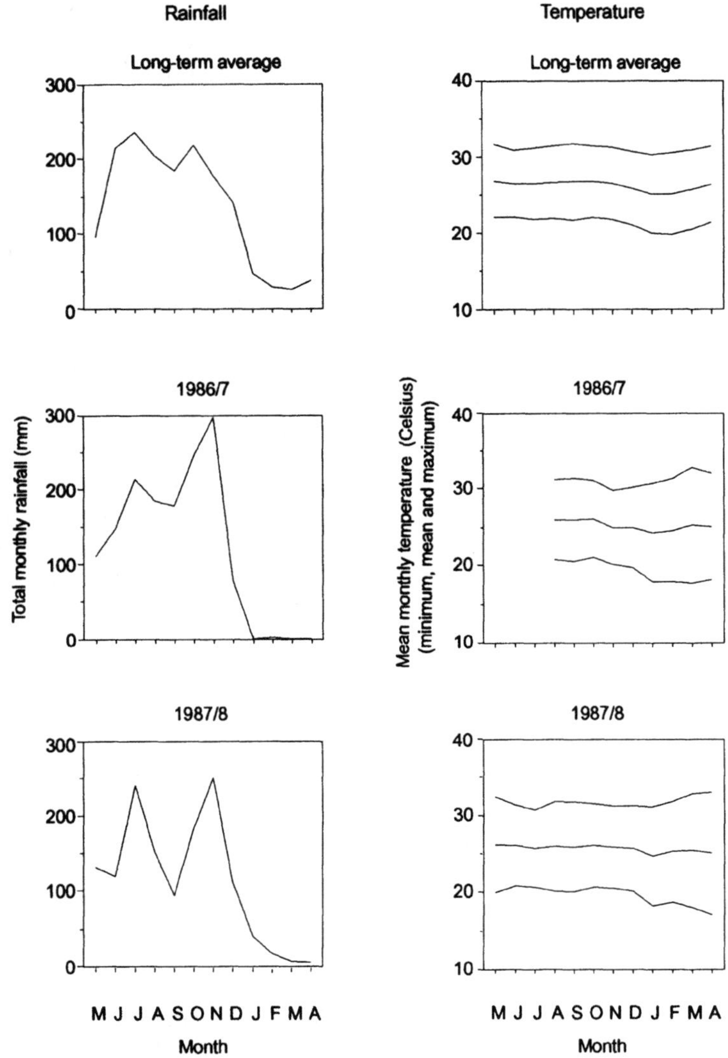

Figure 9.1. Rainfall and temperature data compared with long-term averages from the Trinidad Site.

majority of pilei in both 1986 and 1987 were 5–15 mm in diameter.

Suspended vegetative brooms

Observations started in week 27 of 1986 on the first set of brooms. The first basidiocarps were observed in week 33 of 1986, and only small amounts were produced until week 40. Intense production then continued for 9 weeks, when it ceased abruptly in week 49 (Figure 9.2). Half the brooms had produced at least one basidiocarp by week 42, and 192 of the initial sample of 200 brooms fructified during 1986 (Table 9.1). Individual brooms were observed with basidiocarps in an average of 7.5 weeks, and the total number of basidiocarps produced on suspended vegetative brooms was 3,741, or an average of 19.5 per productive broom. A highly significant correlation for numbers of basidiocarps observed was obtained between suspended vegetative brooms in the upper and mid canopy (r = 0.98).

TABLE 9.1
Principal descriptors of basidiocarp production on suspended vegetative broom sets and marked diseased cushions, Trinidad site.

Variable	Vegetative brooms			Diseased cushions	
	Set 1		Set 2		
	1986	1987	1987	1986	1987
% productive objects[1]	96	66	42	34	13
Mean no. b'carps /prod. object	19.5	19.2	10.6	5.7	4.8
Mean no. weeks with b'carps	7.5	6.8	3.8	3.7	4.9

[1] brooms or cushions.

Basidiocarp production stopped at the end of week 49 in 1986, and none were observed during the dry season in 1987 (weeks 1–20). Production resumed in week 26, peaked in week 45, and continued to week 52 (Figure 9.3). The total number of basidiocarps produced, proportion of Productive brooms, and frequency of activity all decreased in 1987 (Table 9.1). Average numbers of basidiocarps per productive broom did not change significantly. A total of 123 brooms out of 200 decomposed sufficiently during 1987 to fall from the strings suspending them, 40 of these during the dry season. Of the other 83, 18 fell in early 1988 without producing basidiocarps.

Set 2 suspended vegetative brooms produced less basidiocarps than the Set 1 in 1986 or 1987 (Table 9.1). Basidiocarps formed between week 33 and week 51 of 1987, with a clear peak in production in week 45 (Figure 9.3). A high correlation (r = 0.91) was again obtained between numbers of basidiocarps on brooms suspended in the upper canopy and brooms in the lower part of the canopy.

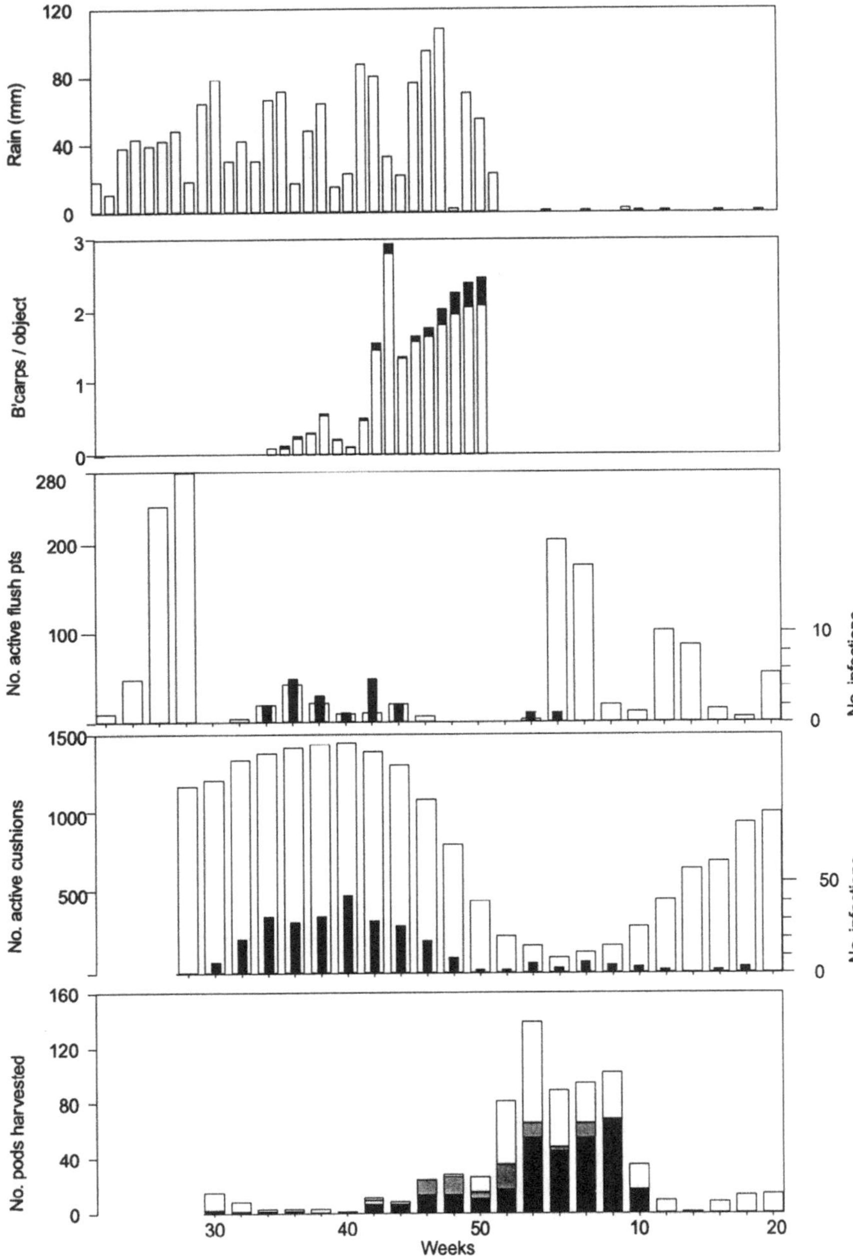

Figure 9.2. Rainfall, basidiocarp production[1], vegetative flushing and infection[2], flower cushion activity and infection[3], fruiting and fruit infection[4] in the Trinidad Site in 1986/7.

[1] open columns, the most productive source type and closed columns, sum of the rest; [2] open columns, no. of healthy flushes and closed columns, no. of infections; [3] open columns, no. of active healthy cushions and closed columns, no. of infections; [4] closed columns, witches' broom infected, shaded columns, other diseases, and open columns, mature healthy fruit.

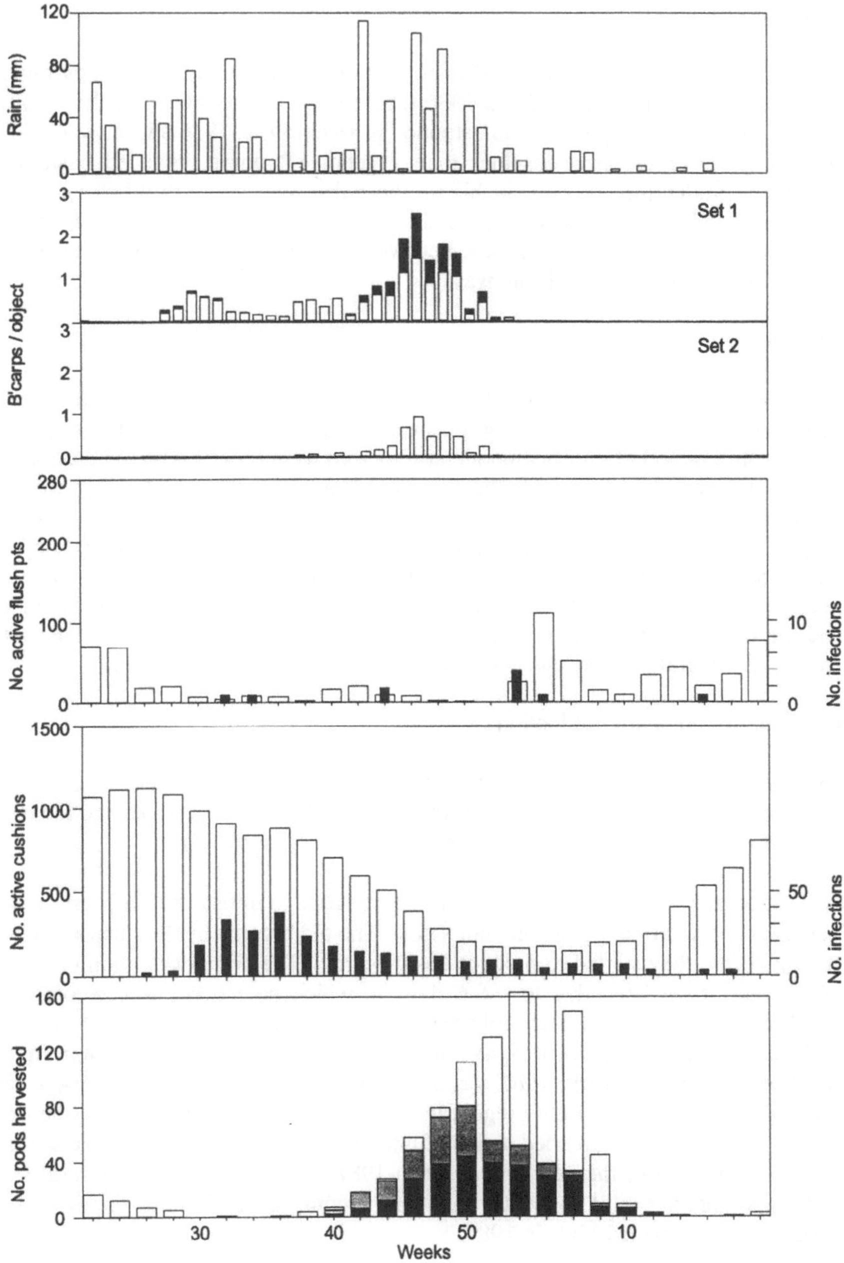

Figure 9.3. Rainfall, basidiocarp production[1], vegetative flushing and infection[2], flower cushion activity and infection[3], fruiting and fruit infection[4] in the Trinidad Site in 1987/8.

[1] open columns, the most productive source type and closed columns, sum of the rest; [2] open columns, no. of healthy flushes and closed columns, no. of infections; [3] open columns, no. of active healthy cushions and closed columns, no. of infections; [4] closed columns, witches' broom infected, shaded columns, other diseases, and open columns, mature healthy fruit.

Cushion brooms

Basidiocarps were found from week 25 in both years on the marked diseased cushions, though so few were found up to week 34 in 1986 that the data do not show on the graph (Figures 9.2 and 9.3). Basidiocarp production stopped in week 49 of 1986 and in week 50 of 1987. Significantly fewer diseased cushions than vegetative brooms produced basidiocarps, but frequency of production was comparable for those that were active (Table 9.1). Basidiocarps were never found on cushions with infected flowers alone, the commonest type in the sample. Eighty per cent of all basidiocarps counted were found on cushions with brooms, with the other 20% on cushions with chirimoyas.

Other sources

Only one suspended witches' broom pod out of 50 produced any basidiocarps in 1986, perhaps partly because pods were only selected late in both years, and two pods in 1987. Over the entire period of observation no basidiocarps were observed on the broom or pod sets placed on the ground as directed in the Protocol. However a few basidiocarps were produced on a separate sample of 50 brooms pushed into the soil so that they stood vertically (18 sporophores, 8 active brooms). Basidiocarps were also occasionally observed on brooms on the ground in other parts of the same plantation, but only on those not covered by leaf litter. Where detached brooms had been placed in heaps, some basidiocarp production was noted at the top of the heap. Brooms placed on the ground at the beginning of the rainy season normally disintegrated completely by the end of the year (28–36 weeks).

Climate and basidiocarp production

Rainfall started in week 20 of 1986 and continued up to week 50 at an average of 47 mm per week (Figure 9.2). It started again in week 20 of 1987 but continued beyond the end of 1987 to week 4 of 1988 at an average of 39 mm per week (Figure 9.3).

Approximately 13 weeks with rain were required to initiate fruiting in the new samples of suspended vegetative brooms in both years. The old brooms in 1987 needed only 5 weeks with rain before they started to produce basidiocarps. The abrupt cessation of production in week 49 may have been due to the 6 days in which no rainfall was recorded from 6th December. The rain on the 12th, 13th and 14th December (total of 23 mm) in week 50 was insufficient to induce basidiocarps, and almost no further rain fell until 2 mm in week 9 of 1987. The dry period in early 1987 was exceptionally severe, as only 6.5 mm of rain fell in 22 weeks, with relative humidity regularly less than 50% and temperatures often over 33°C in March and April. The extended period of unfavourable conditions may have caused the less abundant production of the new broom set in 1987, while sufficient biomass of saprophytic mycelium might have built up already in the old brooms from 1986 that their total production of basidiocarps remained high in 1987.

The periods of basidiocarp production for 1986 and 1987 differed in the amount and distribution of rainfall. More basidiocarps were produced in 1986, which had a total of 1,458 mm of rain, of which 955 mm fell after week 32, when most production occurred. There were 55 days with more than 5 mm of rain. Recorded values for all of these variables were lower in 1987, which also may have reduced production on the new sample of brooms. For both years during the periods of basidiocarp production, the average minimum and maximum temperatures ranged between 20–22°C and 30–32°C, respectively. Maximum relative humidity was usually around 95–96%, with the minimum varying between 60 and 70%.

Relationships between numbers of basidiocarps and climate were analysed by both simple and multiple linear regression analysis. The total number of basidiocarps on Set 1 was used as the dependent variable for 1986 and 1987, against a range of ten independent variables. These were rain total and frequency; mean temperature and duration above 30°C and below 20°C; mean relative humidity and duration above 95% and below 70%; and frequency of wetness after midday. For both years there was a positive relationship between rainfall and basidiocarp production but it was not significant. Results obtained from the other climatic parameters were not significant and were inconclusive. It was clear that production was promoted by well distributed rain as observed in 1986. However, explanation of basidiocarp numbers on a week to week basis was not possible.

Flushing and Vegetative Infection

Flushing

During 1986 there were three main and two subsidiary periods of shoot activity with the greatest at the beginning of the experiment in week 25/27 (Figure 9.2). A total of 706 active growing points was recorded from 16 observations. In 1987 there were 5 distinct periods of vegetative activity, with three main peaks occurring during the first half of the year. The highest numbers of growing points and new leaves were noted during weeks 3, 5 and 11 (Figure 9.3). Vegetative growth for the first 19 weeks in 1988 followed the 1987 pattern but with fewer active points per flush period.

The degree of correlation of vegetative growth from tree to tree was evaluated using the pooled data for the total number of growing points counted once per fortnight for the five terminal units per tree during the course of the experiment ($n = 52$). All the trees showed good correlation at the 5% level, except two coefficients which were not significant. The five TSH 1188 trees (nos 1–5) were significantly correlated at the 1% level. Trees 1 and 3 were found to have the greatest number of growing points of all ten trees, and trees 2 and 5 the least. The mean number of leaves per active growing point ranged from 3.8 to 4.7.

Vegetative infection

The largest number of active growing points occurred in week 27 1986, just coinciding with the first observations of basidiocarps on cushions, after which production remained sparse until week 33 (see above). The total number of growing points was low throughout the rest of the period in 1986 when basidiocarp numbers were high (Figure 9.2). The sum of active point counts in weeks when basidiocarps were observed in 1986 was 408 of which 279 were before week 30. Vegetative infections were observed from week 31 of 1986, and were most likely initiated during the intense flush of weeks 25/27. Twenty separate infection points were found within the branch units, 13 of them on tree 1. Vegetative activity on that tree was similar to the other nine. A total of 15 brooms formed directly or indirectly from the 20 infections, four of which (axillary) formed from swollen stem and pulvinus/petiole infections approximately 4–6 weeks after the first symptoms were observed. Cankers were also noted on chupons and young branches outside the terminal branch units, but these were rare and difficult to detect.

Vegetative shoot growth in 1987 occurred predominantly during the early part of the year from week 1 to week 23, during which time no basidiocarps were observed. The number of basidiocarps recorded before week 35 in 1987 was higher than in the previous year due to the old brooms from 1986. A total of 124 active points was recorded from week 25 to 51 when basidiocarps were present, and early infections appeared in weeks 31 and 33. These may have been initiated in the small amount of flushing in weeks 25 and 27. Ten brooms formed directly from 10 separate infection points, with no other classes of infection. No single tree had a large proportion of infection, and trees 6 and 7 (TSH 919) showed no shoot activity in the inoculum period and had no infection.

A disease index was calculated as "sum of infections/sum of active point counts", and pooled data from the ten trees indicated that infection on the vegetative parts of the terminal units increased from 0.04 in 1986/87 to 0.08 in 1987/88. An analysis of variance showed that the difference was not significant. The index for 1986/7 was greatly influenced by the small amount of infection on the intense early flush, which itself might have been due to the very small number of basidiocarps observed at that time. The index for 1986/7 would have been appreciably higher had this flush not been included. In 1987/8 the total number of vegetative infections in the units fell from 20 to 10 and on the ten experimental trees decreased from 147 to 104. Sixty four per cent of all the brooms formed on the five TSH 1188 trees.

Flowering and Cushion Infection

Flowering

A clear annual pattern was found in cushion activity for the 24 months of the experiment. The highest numbers of active cushions recorded in each year were 1,450 in week 39 of 1986, and 1,122 in week 25 of 1987. Active cushion counts decreased after the peak of flowering toward the end of the year, reaching a minimum around week 3 of the following year in both cases (Figures 9.2 and 9.3). The greatest number of active cushions was recorded on tree 8 and the lowest on tree 10, both belonging to the TSH 919 clone. There was less variation in the total number of cushions on the marked branches of the five trees of TSH 1188 than the TSH 919 trees. The mean number of active cushions per metre of branch was 15 for TSH 1188, compared with 12 for TSH 919.

Numbers of cushions with open flowers were highly correlated with numbers of active cushions ($r = 0.93$). The mean number of cushions with open flowers per branch unit per fortnight for the TSH 1188 clone ranged from 1.6 to 1.9, and from 1.0 to 2.2 for the TSH 919 clone. Synchronization of cushion activity between trees was evaluated by correlation analysis, using the pooled data per tree per fortnight for the duration of the investigation. Results indicated that all ten trees were highly correlated, irrespective of clone.

Cushion infection

Infections on cushions were placed into the different classes as laid out in the Protocol. The majority of new infections appeared as infected flowers only (Class 1), 76 and 78% of all records in 1986 and 1987 respectively, and the predominant symptoms were unabscissed necrotic flowers, often with swollen pedicels. Chirimoyas were produced on 17.5% of all infected cushions but very few formed on trees of the TSH 1188 clone. A total of 73 cushion brooms was observed on the marked branch units, of which 42 (58%) formed on the TSH 919 trees. Brooms on TSH 919 had a mean length of 25 cm, compared with 12 cm for brooms on TSH 1188.

Whole tree counts found that on the five TSH 919 trees there were 225 infections with vegetative cushion brooms compared with 131 in the 1188 clone. The majority of brooms counted were observed on the upper canopy and were smaller than those observed in the marked units on the larger lower branches.

The different symptoms on infected cushions were not produced in ordered sequences or preferentially at different times of year. However, most vegetative cushion brooms (89%) on the marked branches formed on cushions that had previously shown some symptom of infection. Cushions were observed producing both healthy and diseased flowers, and healthy pods were seen on cushions that were obviously infected.

In 1986, infected cushion sets were already producing basidiocarps in week 27, the first records of infection on cushions in 1986 occurred in week 29. Incidence of new infections reached a maximum of 42 in week 39 (Figure 9.2). In 1987, the first symptoms of infection were observed in week 27, and the largest number of new records (28) occurred in week 35 (Figure 9.3). The steady decline in numbers of new infections after the peaks in both years might have been related to reduced flowering. Few new infections were recorded outside inoculum periods in either year.

Most basidiocarp production of *Crinipellis perniciosa* on suspended vegetative brooms occurred during the latter part of the production period in both years, as the number of cushion infections was decreasing (Figures 9.2 and 9.3). Large proportions of infection in both years (61 and 62%) appeared before week 40 when basidiocarps were relatively scarce. The proportions increased to 92 and 81% if a 6 week incubation period was allowed from week 40. The greatest number of new infections were found on tree 7 (TSH 919) for both years, but the analysis of variance for a difference in infection between the two groups of five trees of each clone was not significant. The mean number of new infections per tree was 22.4 in 1986 and 18.1 in 1987, and both average cushion activity and numbers of cushions with open flowers were lower in 1987 (831 and 108) than 1986 (1448 and 172).

A disease index was calculated for each tree as "sum of infections/ sum of infections + sum of healthy cushion counts" for both active cushions and cushions with open flowers. Pooled data from all ten trees showed that the two indices did not increase significantly from 1986/7 to 1987/8 (Table 9.2). Indices for TSH 919 were higher than TSH 1188. Variation in cushion infection between trees was greater for TSH 1188 than TSH 919, as shown by correlation coefficients between sums of healthy cushion counts and sums of infections. Coefficients were only significant for TSH 919, and for pooled data for all ten trees were only significant for sums of cushions with open flowers.

TABLE 9.2
Cushion disease indices and correlation coefficients for relationships between sums of healthy cushion counts and sums of infections for the clones TSH 1188 and TSH 919, Trinidad site.

Year	Disease	Active cushions			Cushions with open flowers		
		1188	919	Combined	1188	919	Combined
1986/7	index	.064	.121	.092	.398	.504	.460
	r	.000	.402	.176	.412	.809*	.729*
1987/8	index	.110	.143	.126	.514	.540	.526
	r	.595	.964*	.582	.680	.867*	.771**

*, ** indicates coefficients significant at $P < 0.05$ and 0.01.

Fruiting and Fruit Infection

Fruit set began in week 27 in 1986 and continued until week 49 with a slight peak from week 31 to 39. Fruit set in 1987 was almost entirely concentrated into the period weeks 31 to 39. Total pod production for the 50 site trees was significantly correlated between the two years ($r = 0.61$). The majority of pods completed the first 12 weeks of their development, during which they were susceptible to infection by witches' broom, within the period when basidiocarps were observed on marked sources. Harvests of healthy mature pods or necrotic pods started around week 40 and continued to week 10 in both years. Few witches' broom diseased pods were recorded outside the periods of basidiocarp production, allowing for a 10–12 week interval for incubation between infection and appearance of symptoms.

The greatest incidence of pods diseased with witches' broom occurred 10–12 weeks after the main peak of basidiocarp production in both years. Allowing for the incubation period of 12 weeks, there was a significant positive correlation between the two sets of data in 1986 ($r = 0.73$), but not in 1987 ($r = 0.08$). This relationship was not improved by using the fraction of available pods that became diseased, calculated as the total of pods present less pods already diseased or no longer susceptible (over 12 weeks old). The index related to available pods could only be calculated for fruit set within the marked branch units where fruit ages were known, and accuracy was reduced by small sample sizes. There was no relationship between percentage witches' broom pods on the 50 site trees between the two years. Significantly more pods became diseased from TSH 919 than TSH 1188; 50% versus 38% in 1986/7, and 31% versus 19% in 1987/8.

Pods infected by *Phytophthora* pod rot were also recorded throughout the harvest periods; comprising about 10% of all pods in 1986/7 and 17% in 1987/8. The proportion increased to 21 and 28% respectively, if only the marked branches were considered, because the marked units were located in the lower branches of the tree. All phytophthora–infected pods were found at or below 2 m above the ground, and almost all pod disease below 1 m was caused by that pathogen. Most pods formed between 1 and 3 m above the ground, and pods diseased with witches' broom were commonest in that region. A vertical disease gradient was clear for *Phytophthora* pod rot, and this affected the spatial distribution of witches' broom so that pod rot incidence of the latter disease was apparently reduced nearer the ground.

TABLE 9.3
Comparison between different samples for pod harvest records in 1986/7 and 1987/8, Trinidad site.

Year	Sample size	Total harvested pods	% witches' broom	% *Phytoph–thora*
1986/7	50 units	66	25.8	21.2
	10 trees	213	31.9	11.7
	50 trees	779	39.2	10.0
1987/8	50 units	95	28.4	28.4
	10 trees	246	30.1	17.1
	50 trees	983	28.6	16.7

Proportions of pods in each class were distorted if only the marked branches were considered, but accuracy was not necessarily improved when all 50 trees were considered against the 10 main site trees (Table 9.3). Between 30 and 40% of pods became diseased with witches' broom in 1986/7, and 29 to 30% in 1987/8. The proportion of witches' broom pods still decreased in 1987/8 if pods with *Phytophthora* rot were removed from the total.

Comparison with Epidemiological Studies in Trinidad, 1938–1942

One previous study in Trinidad, as published by Baker & Crowdy (1943, 1944), merits special comparison with the data collected by IWBP. Amongst other subjects, the first paper presents detailed data on basidiocarp production in two localities (called here 'A' and 'B'), over almost 2 calendar years to early 1942. The 'A' site was located in a low rainfall area, and 'B' in a high rainfall area. The second paper reports on a comprehensive study of: rainfall, sporulation, shoot and flower cushion activity with associated infection, and some aspects of pod disease. This work was carried out in a plot of 76 trees of open pollinated ICS 1 progeny at a different site (called here 'C') between September 1939 and September 1942.

Basidiocarp production

The IWBP work supports nearly all the conclusions from the studies of Baker and Crowdy between 1940 and 1942. The proportion of brooms producing basidiocarps and the mean number per broom were similar: basidiocarps formed on 60–75% of vegetative brooms, which produced a mean of approximately 20 basidiocarps in 1942. They reported an induction period without basidiocarp production after broom necrosis, followed by continuous sporulation except at the height of the dry seasons.

The scarcity of dry–season rain varied in the six site/year combinations studied by Baker and Crowdy, but no basidiocarps formed in either year at site 'A' for at least 8 weeks. No such clear break in production was recorded in either year at either site 'B' or 'C', partly due to the absence of a significant period with little or no rain as occurred in 1987 in Trinidad. Early rainy season basidiocarp production occurred on older brooms for the earlier work, as it did in the IWBP study. A general relationship between basidiocarp production and amount of rain was also observed by Baker and Crowdy, but correlation analysis of weekly data produced no significant coefficient in either site 'A' or 'B'.

Similarly, very few basidiocarps formed on brooms in direct contact with the ground in either work, and Baker and Crowdy also reported that brooms fructified at the tops of heaps. Mummified pods produced very little in their first year, and not at all in the second due to decomposition.

Phenology of growth and infection

The data of Baker and Crowdy showed fundamental differences in phenology between the trees they studied, and the TSH trees used by IWBP which are now widely planted in Trinidad. The ICS 1 progeny flushed six or seven times throughout the year, at 5–7 week intervals. The heavy flushes in the second half of the year during the wettest season did not occur on the TSH materials. Baker and Crowdy noted that brooms appeared in two distinguishable periods, most in January/February and some in August–October. The severely reduced broom formation by the TSH trees in January is probably due to reduced flushing in November compared with the ICS 1 progeny. A mean of 135 vegetative brooms/tree was observed by Baker and Crowdy in 1942 and only 19 brooms/tree in the TSH plots at the start of IWBP.

The ICS 1 progeny were shown to flower throughout the year, without the distinct reduction in cushion activity towards the end of the rainy season found in the TSH trees. A mean of 70 cushion brooms was found on the former in 1942, compared with 18 cushion brooms on the TSH trees in 1986. Baker and Crowdy did report a highly significant correlation between the number of cushion brooms and flower production/cushion number per tree, which was also partly true for the IWBP material. Reduced cushion disease incidence in the IWBP study might be attributable to less cushion activity in the sporulation period.

Diseased pods formed at similar times of year in the two studies. Losses fluctuated at the Baker and Crowdy site, and the published data were insufficient to make a direct comparison with IWBP. Incomplete data for 1942 show 57% of pods diseased with witches' broom, and 16% to other diseases (mostly *Phytophthora* pod rot). Trees in both cases were producing an average of about 20 pods per year each.

CONCLUSIONS

1. The wet season cycle divides the year into two distinct periods. Infection occurred from July to December in the mid–late rainy season, and did not occur from January to June in the dry and early rainy season. No basidiocarp production was recorded from weeks 3 to 24 of either year, and little infection could definitely be traced to that period.

2. Patterns of host growth were consistent between the two years and within the ten site trees for shoot growth, flowering and pod production. Most shoot growth occurred outside the inoculum period except for the May flush which just overlapped the first weeks of

basidiocarp production. Flowering was intense from May to October, and most pod set occurred in August and September.

3. Formation of basidiocarps was induced approximately 13 weeks after the start of the rainy season on brooms formed in the previous year. This period decreased to 5 weeks on older vegetative brooms and diseased cushions. Vegetative brooms were the more productive of the two sources. Few sporophores were produced on diseased pods in the canopy, and none at all on any objects on the ground provided they were in complete contact with it and/or covered with leaf–litter or other material.

4. Early basidiocarp production from June (week 26) to late August (week 35) occurred only on infected cushions (brooms and chirimoyas) in 1986 and 1987, and old vegetative material from the first year in 1987. Most vegetative and flower cushion infection was initiated during the period up to week 40, so the older sources of basidiocarps were particularly important in shoot and flower infection. Disease indices in 1987/8 were not significantly different from those for 1986/7 for either type of tissue. Amounts of cushion infection were loosely related to amounts of cushion activity.

5. Maximum basidiocarp production occurred between October (week 40) and the end of December (week 52). Most pods were infected during this period, but incidence of vegetative and cushion infection was relatively low due to reduced meristematic activity. The proportion of pods diseased with witches' broom decreased from 1986/7 to 1987/8, even if adjustments were made for the greater incidence of *Phytophthora* pod rot in the second year. The decrease might have been due to reduced numbers of basidiocarps or less frequent infection periods. Changes in numbers of witches' broom diseased pods were related to changes in basidiocarp numbers in 1986, allowing for a 10–12 week incubation.

MANAGEMENT IMPLICATIONS

1. Phytosanitation would have to include a main removal of brooms between March (week 10) and May (week 20), with perhaps a second removal before September (week 35). The first removal would reduce the limited basidiocarp production in the early part of the season, and the second removal would include the late–emerging or missed brooms before the main period of fructification. Removed brooms could be left on the ground within the plantation if they did not project above the litter layer or were not left in large piles. Reduction in disease could become more effective with continued application over several years, as the older diseased material is primarily responsible for new vegetative and cushion infections. The TSH materials show less shoot activity in the inoculum period than the ICS materials investigated by Baker & Crowdy (1944), but cushion infection is still a significant problem in the Santa Cruz area.

2. Chemicals could be used in a variety of ways. Protection of developing pods using inorganic compounds (coppers) could be achieved by sprays in the peak infection period October to mid December (weeks 40–50). This would require a maximum of five sprays even on a 15 day cycle. Systemic compounds for protection of vegetative or flower cushion activity would be best applied in May at the end of the dry season (week 16–18). Eradicant

compounds used to kill developing mycelium in green brooms could be applied slightly later in July, as most diseased tissue appears from August to November (weeks 30–48). Soil applications might have to be made slightly earlier, to allow time for the chemical to be translocated to growing apices.

Chapter 10

COMPARATIVE EPIDEMIOLOGY STUDY: VENEZUELA

R. Silva–Acuña, E.H. Reyes & R.F. Colmenares

INTRODUCTION

Importance of Cocoa and Witches' Broom in Venezuela

Cocoa played an important role in the social and economic life of the country, although petroleum has dominated recent development. During the colonial period (1600–1800) cultivation of cocoa spread through the country, and it became the principal export product. In the Republican era after independence it continued to occupy an important place among exports. Today, there are three main regions of cocoa production: the central region with 30,000 ha, the eastern with 29,000 ha, and the western with 6,000 ha. These produce 15,000 t of dry cocoa (Capriles, 1988) and directly or indirectly generate 40,000 jobs. Cocoa is still one of the principal export products.

Witches' broom was first reported by Singh (1937) in cocoa plantations in the east of the country in Delta Amacuro, and that same year Posnette & Palma (1944) recorded it in plantations in the nearby Paria Peninsula. The disease was first recorded in Trinidad in 1928, although there was no evidence that the two outbreaks were linked. It was found later in Miranda State in 1945 (Hernández, 1958), in the central region with good potential for the development of cocoa. The Ministry of Agriculture and Animal Husbandry took steps to control the disease, and in 1958 the Miranda Experimental Station was created in this region with a prime objective as the study of witches' broom disease and research into cocoa agronomy.

An outbreak of witches' broom disease in 1970 in the south–western states of Táchira and Apure, a region bordering Colombia, caused the eventual abandonment of 3,000 ha of cocoa (Capriles & Reyes, 1962). Pod losses varied in the above states (Capriles, 1973), apparently dependent on local climatic conditions. The IWBP site was located in Táchira State in a small area of cocoa cultivation on the Isla de Betancourt which had totalled about 100 ha in 1975. Witches' broom was first recorded in 1978, and was causing pod losses of over 50% after only a few years. Farmers began to replace cocoa with other crops, and the remaining 20 ha is only viable due to biannual broom removal. Cocoa in other areas of the country was less severely affected by the disease, and the states of Portuguesa, Merida and the southern zone south of Lake Maracaibo remains free of the disease (Capriles, 1977).

Disease Management in Cocoa: Comparative epidemiology of witches' broom. Edited by S.A. Rudgard, A.C. Maddison and T. Andebrhan. Published in 1993 by Chapman & Hall, London. ISBN 0 412 58190 6

Witches' Broom Research in Venezuela

Most cocoa growers in Miranda State in the 1950s were supplied with vegetatively propagated cuttings selected for their high yield and large beans from clones such as SC6 and UF 667 or UF 676. They turned out to be highly susceptible to witches' broom (Hernández, 1958), and a campaign to eliminate such trees was begun in 1963. Regional trials were started for the evaluation of hybrid material from known crosses (Parra, 1989), and reaction to witches' broom was an important criterion where the disease was serious (Reyes, 1963). Several resistant crosses were identified after programmes of up to 10 years. Susceptible materials such as GS were also identified and eliminated from the selection programmes (Capriles & Rojas, 1977). Particular attention was paid to progeny of Sca 6, which showed tendencies to bear early, yield well and be resistant to witches' broom (Pérez *et al.*, 1972). The beans from such crosses tended to be small, so they were only planted in areas of high disease incidence. Recent advances in the breeding programme (Reyes, 1973) have included the selection of progeny from crosses between Porcelana and Amazon types, which have not shown any symptoms of witches' broom attack over 5 years.

A comparative study on the production of basidiocarps was carried out (Capriles, 1983) using samples of brooms from different parts of the country, subjected to a standard wet and dry regime (16 h with and 8 h without wetting, respectively). The timing of emergence varied greatly as well as the number of basidiocarps formed. Isolates of vegetative mycelium of *Crinipellis perniciosa* (Materán, 1981) from different cocoa regions grown on artificial media showed variation in form, colouring and growth rate when maintained in the same conditions. Epidemiological studies conducted by the Miranda Experimental Station (Capriles, 1973) revealed a general positive relationship between the formation of new vegetative brooms and rainfall.

Studies in the laboratory, greenhouse and field, have shown that several systemic fungicides (Capriles, 1983) inhibit the formation of basidiocarps when sprayed on to necrotic brooms. One of the above, tridemorph, was sprayed at intervals of 30 days on mature trees (Capriles, 1988), resulting in reduced broom formation, as well as a significant increase in cocoa production. Protectant compounds such as triphenyl tin acetate and copper oxychloride reduced pod losses when sprayed on to trees without inhibiting the formation of vegetative brooms (Capriles & Freites, 1977).

Current Practices for Disease Management of Witches' Broom

At present, there exist recommendations (Parra, 1989) for general management of the plantation as well as witches' broom disease, which are as follows:

- Rationalisation of shade by thinning;
- Opening of drains to eliminate excess ground water;
- Sanitary pruning twice a year during the dry seasons or times of less rain (March–April and September–October);
- Collection of diseased fruits and harvest waste, which should be eliminated from the plantation;
- Eradication of affected trees and planting of material of proven tolerance to the disease;

- Protection of the crop with fungicides during the first three months of the period of greatest fruit formation.

MATERIALS AND METHODS

Description of the Experimental Site at El Piñal

The site was located in an isolated cocoa plantation 11 km from the town of El Piñal at La Primavera farm on the Isla de Betancourt. The farm had 5 ha of seedling criollo cocoa of different ages originating from Ocumare de la Costa, Aragua State. The planting density was 625 trees/ha with a distance of 4 x 4 m, with irregular shade of *Erythrina* sp. planted at about 7 x 7 m. The soil was a silty clay loam with low potassium and organic matter status, medium status for calcium, phosphorus and nitrates, a pH of 5.3, and a high moisture retention capacity. The whole area had drainage channels which facilitated the removal of surface water.

The farmer normally removed weeds and brooms together twice a year, first before the beginning of the rains (April) and then at the end of the rains (December). The latter was combined with a formation pruning to eliminate undesirable branches or to shape the canopy. The brooms removed from the trees were scattered on the ground in the plantation.

Site Installation

A block of 50 trees for the experimental area was selected from the plantation. Recording began in week 30 of 1987 and continued until week 52 of 1989, when the study finished. The 10 sample trees were selected at random from the experimental area, and units of the trunk and branches were marked according to the protocol described in Chapter 4. Records of pod harvests were kept for the other 40 trees in the experimental area.

No diseased material was removed from the site trees when the experiment was installed. Infections of all types were sparse, as brooms had been removed consistently by the farmer. Brooms and *Crinipellis*-infected pods for the first set (Set 1) of objects for observation of basidiocarp production had to be gathered from a nearby plantation. Evaluation of sporulation in cushion brooms was not recorded in 1987 and 1988 as none existed in the plantation. The brooms placed on the ground in 1988 rapidly decomposed (no sporulation being recorded in that year). Additional sets of 40 brooms each were put in 1989 in two places in the experimental area on trays of plastic mesh of different sizes, supported by a metal frame 20 cm above the ground.

The meteorological station was installed in a clear and flat area adjacent to the trial at a distance of 60 m. Temperature and rainfall data were also obtained for the 7 years before the experiment from the meteorological station of the airport of Santo Domingo, 5 km away from the plantation.

RESULTS AND DISCUSSION

Climate

Rainfall

At El Piñal there was generally a defined period of reduced rainfall in January–March. The rainy season was from April until December. The months May to October of the wet season had the most rain, with June being the wettest month (Figure 10.1). Rainfall normally decreased in November–December. Figure 10.1 presents a comparison between mean rainfall data over 7 years from the Santo Domingo meteorological station (of the Venezuelan Air Force), with the site meteorology station with 2 years' data. The major seasonal variations in rainfall were maintained in the area under study. Differences between the months of highest and lowest rainfall, particularly in 1989 (June–September) were within previously observed limits.

For the period under study, dry season (January–February) rainfall was 16 mm in 1988 and 171 mm in 1989. During the rainy season, beginning in April 1988, a gradual increase in rainfall was observed up to June when 905 mm were recorded, the highest monthly total for the year (Figure 10.1). Less rain fell in August and September, but with great intensity. From October the rains diminished until December when 156 mm were recorded. Total rainfall for the year was 3,096 mm. The rains began intermittently in 1989 as in the previous year, before increasing in May. July was the wettest month with 397 mm, and the amount of rain decreased gradually, as in 1988, to 12 mm in December. Total rainfall for 1989 was 2,253 mm.

Temperature

Maximum and minimum temperatures for the two years studied were similar (Figure 10.1). In association with the rain, the temperature during the period of least rain increased gradually from January with the same peak in April for both years, with average values between 28 and 29°C. After that, the temperature decreased in the rainy season and values varied between 24 and 26°C by the end of both years.

Basidiocarp Production

Comparison of sources

The brooms suspended in the upper and middle canopy produced the most basidiocarps, and the pods at these levels produced hardly any (Table 10.1). The pods at soil level produced almost no basidiocarps, and the ground brooms very few. No basidiocarps were formed on the brooms placed on plastic mesh 20 cm above the ground.

The brooms and pods in contact with the ground did not produce many basidiocarps due to the high soil humidity which caused rapid decomposition of the woody tissue of the brooms. Isolating the brooms from immediate contact with the soil did not sufficiently increase drying times so as to increase production. Isolated cushion brooms with basidiocarps

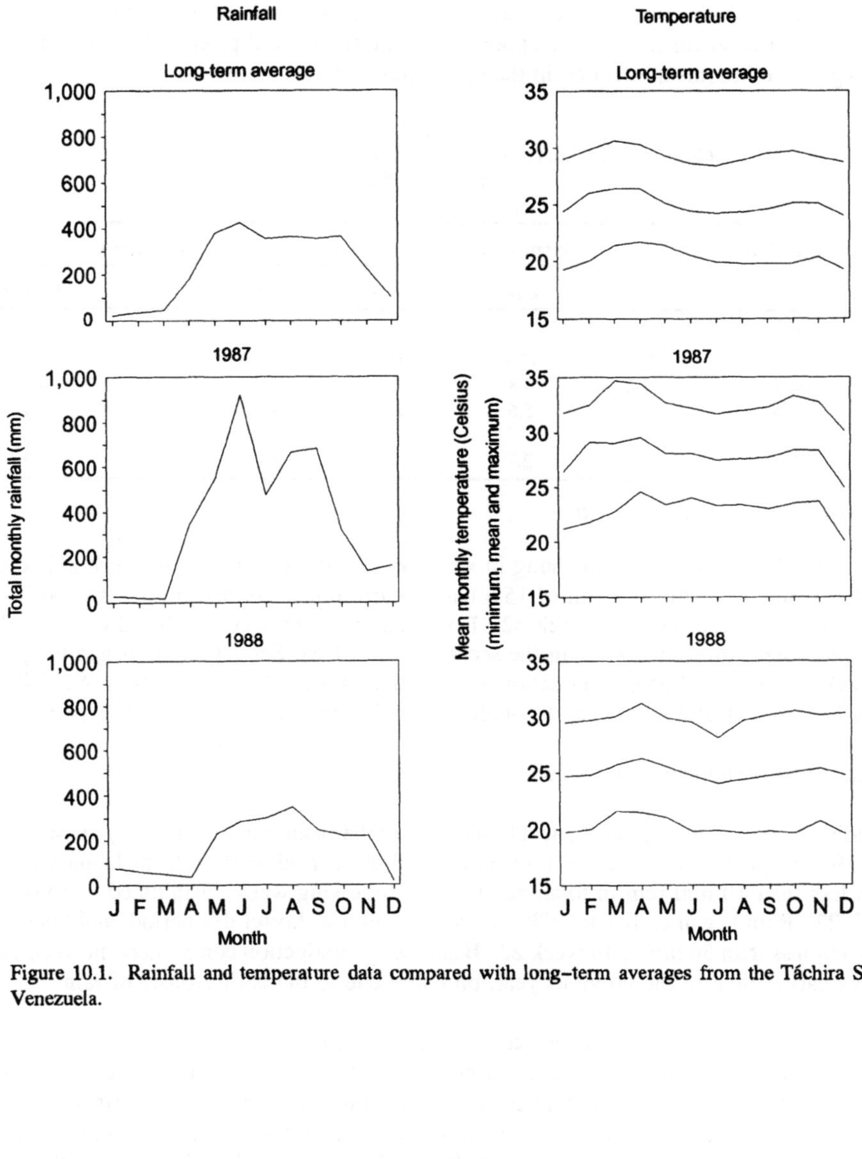

Figure 10.1. Rainfall and temperature data compared with long-term averages from the Táchira Site, Venezuela.

on the trunks and lower canopy outside the units were also rarely observed in 1988 and 1989. Brooms formed on cushions in the upper part of the canopy did produce basidiocarps, as did the vegetative brooms suspended in the upper and middle canopy.

TABLE 10.1
Mean totals of basidiocarps object on various sources in different years at El Piñal

Set & Year	Vegetative brooms			Diseased pods		
	High	Low	Ground	High	Low	Ground
1st Set						
1987[1]	17.3	13.9	0.3	-	-	-
1988	7.2	4.8	0.1	0.4	0.1	0.1
1989	8.2	5.6	0.2	0.6	0.1	0
2nd Set						
1989	5.4	2.7	0.1	0.5	0	0

[1] incomplete year starting in week 30.

Set 1 vegetative brooms hung in week 30 of 1987 began to produce basidiocarps in week 33, and a combined mean of 15.6 basidiocarps/object formed on the two positions in the canopy between then and week 52. Fewer basidiocarps were produced on Set 1 during 1988 and Set 2 hung in 1989. In the second year (1988), Set 1 brooms began to sporulate in week 22, and basidiocarp production continued until week 48 (Figure 10.2). Set 2 brooms, hung in week 14 of 1989, began to produce basidiocarps in week 22 (Figure 10.3).

Climate and basidiocarp production

Basidiocarp formation began 9 and 11 weeks after the beginning of the rainy season in 1988 and 1989 respectively. There was a well defined dry period in 1988 from January to March with less than 10 mm rain in total, and the rain following week 13 initiated basidiocarps in week 22. Rain began earlier in 1989 (in week 9) after a shorter dry period, and total weekly rain was less than 50 mm until week 22. Basidiocarp production commenced in week 20, two weeks earlier than in the previous year, probably due to the earlier onset of rain.

Simple correlation analysis between weekly means of temperature, relative humidity, and rainfall and basidiocarp production showed that rain was the most important climatic factor in determining amounts of intensity of production. The analyses of correlation between production of basidiocarps by vegetative brooms suspended in the upper and middle parts of the canopy and rainfall were significant at $P < 0.01$; $r = 0.95$ and $r = 0.83$ for 1988 and 1989, respectively. There was no significant correlation for basidiocarps on the ground brooms, partly due to the lack of sporulation; $r = 0.06$ and $r = 0.05$. The ground brooms were starting to decompose in their second year, by the end of which very few were left intact.

Flushing and Vegetative Infection

Flushing

Some shoot growth was almost always found throughout the experiment, but there was a

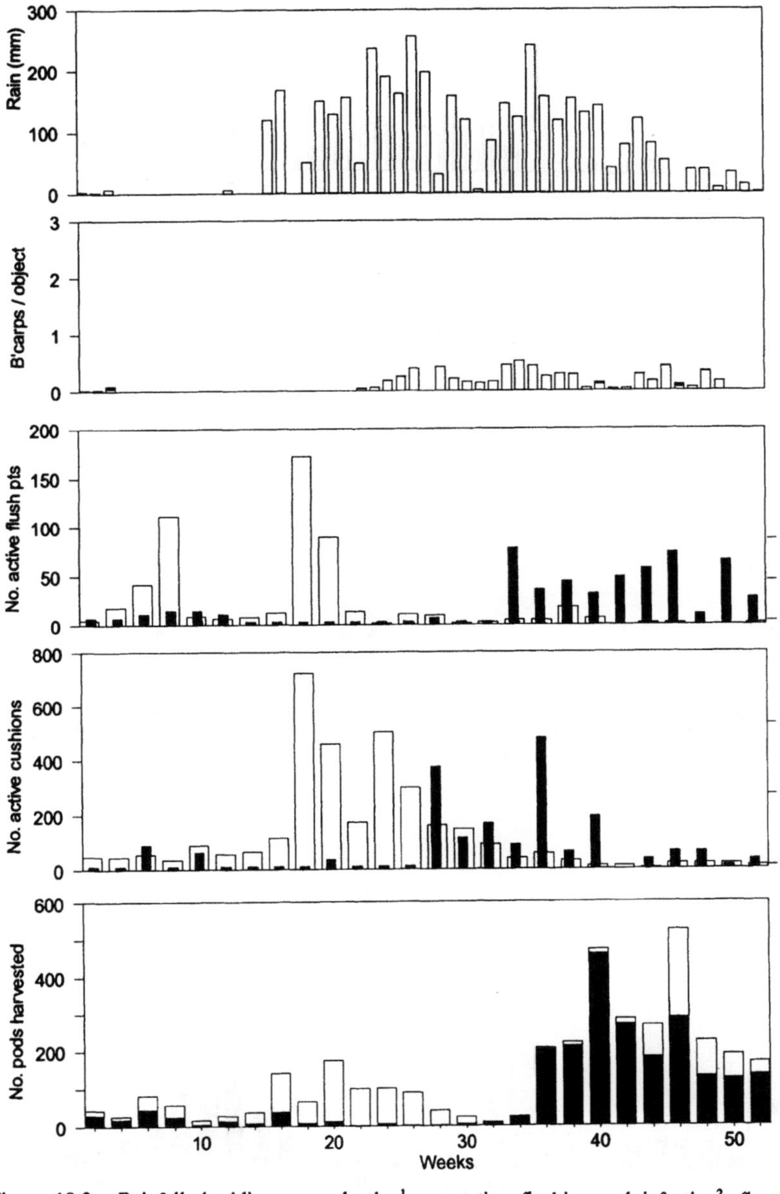

Figure 10.2. Rainfall, basidiocarp production[1], vegetative flushing and infection[2], flower cushion activity and infection[3], fruiting and fruit infection[4] in the Táchira Site, Venezuela in 1987/8.

[1] open columns, the most productive source type and closed columns, sum of the rest; [2] open columns, no. of healthy flushes and closed columns, no. of infections; [3] open columns, no. of active healthy cushions and closed columns, no. of infections; [4] closed columns, witches' broom infected, shaded columns, other diseases, and open columns, mature healthy fruit.

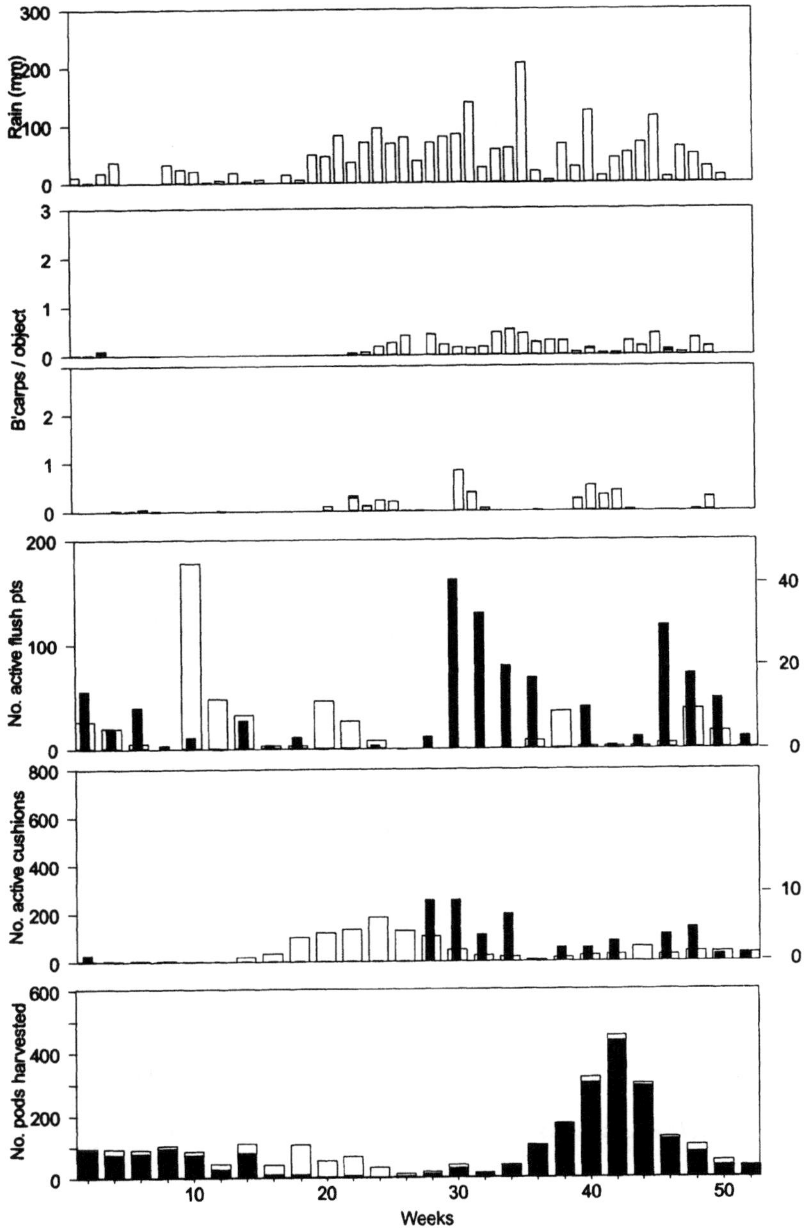

Figure 10.3. Rainfall, basidiocarp production[1], vegetative flushing and infection[2], flower cushion activity and infection[3], fruiting and fruit infection[4] in the Táchira Site, Venezuela in 1988/9.

[1] open columns, the most productive source type and closed columns, sum of the rest; [2] open columns, no. of healthy flushes and closed columns, no. of infections; [3] open columns, no. of active healthy cushions and closed columns, no. of infections; [4] closed columns, witches' broom infected, shaded columns, other diseases, and open columns, mature healthy fruit.

clearly discernible pattern of bursts of bud activity or flushes at clear intervals. Two periods of intense activity were recorded in the first half of 1988; the first began in week 4 and peaked in week 8 with 171 active growing points, and the second larger flush with 282 active points started in week 14 and continued until week 20 (Figure 10.2). The cumulative total of active points during the year was 557, with a mean of 3.0 leaves per flush shoot.

Periods of shoot growth were more frequent, but individually less intense in 1989 (Figure 10.3). The first period and most intense shoot growth occurred between weeks 10 and 14 with 257 active points, and the second between weeks 20 and 22 with 71 active points. Two further minor periods of new growth occurred later in weeks 38 and 48. General flush activity in 1989 was similar to the year before with a cumulative total of 501 active growing points, with a mean of 2.8 leaves per flush shoot.

Vegetative infection

Few diseased shoots were recorded in the new growth in 1987, and they emerged after the main period of shoot growth. The total of new diseased shoots increased from 40 in 1987 to 206 in 1989 (Table 10.2). The greatest proportion of symptoms observed during the study period were brooms, axillary as well as terminal, with swellings very infrequently observed in stems, petioles and pulvini, and no canker lesions observed on stems.

TABLE 10.2
Annual totals of infection types on vegetative growing points at El Piñal

Type of infection	1987	1988	1989	Mean
Swellings				
on leaf bases	1	0	0	0.3
on stems	0	4	0	1.3
Axillary brooms				
with leaf base swellings	0	1	0	0.3
with stem swellings	16	1	1	6.0
without associated swellings	8	74	193	91.7
Terminal brooms	23	39	12	24.7
Total	48	119	206	

Diseased shoots arose in 1988 from weeks 34 to 52 (Figure 10.2). Shoot growth flushes of the crop in 1988 were concentrated between weeks 4 and 20, and the period of greater susceptibility to the pathogen was thus before the period of basidiocarp production. The majority of flush activity also preceded the inoculum period in 1989. Vegetative brooms were formed between weeks 30 and 36 and between 46 and 50 (Figure 10.3). Linear regression analysis showed no relationship in either year between the incidence of vegetative infections and numbers of basidiocarps, allowing for a 6 week incubation period.

A disease index was calculated by dividing the number of new diseased shoots by the cumulative total of active points for the whole year. The values were 0.24 for 1988 and 0.47 for 1989. When a similar index was calculated only using active points observed whilst basidiocarps were being produced, the values were 1.70 and 1.78. The first index increased in 1989 to almost double that of 1988, as had the total number of diseased shoots. Less shoot growth occurred during the period of basidiocarp production in 1988 than in 1989, so the adjustment left the two indexes almost the same. The number of diseased shoots was related to intensity of shoot activity in the inoculum period for the two years of the experiment.

Flowering and Cushion Infection

Flowering

Cushions produced buds and open flowers throughout the whole experiment. Significantly more flowering occurred in 1988 than in 1989, with cumulative totals of active cushions of 2,282 and 917, respectively. Cushion activity was most intense in 1988 between weeks 15 and 29, when 77% of all cushion activity occurred (Figure 10.2), and was most intense between weeks 17 and 27 of 1989, with 64% of all activity (Figure 10.3). Cushions with open flowers were recorded separately and were most numerous during the same periods as the general activity, namely weeks 17 to 23 of 1988 and weeks 19 to 29 of 1989. Fruit set for the main harvest, which took place during the second half of the year, occurred during peak flowering times. Correlation analyses between numbers of active cushions and cushions with open flowers were highly significant (at $P < 0.01$) for 1988 ($r = 0.69$) and for 1989 ($r = 0.78$).

Cushion infection

Diseased cushions mainly arose between weeks 27 and 35, at similar times in the two years. The most frequent types of new symptoms on cushions were flower brooms (Class 1), followed by vegetative cushion brooms (Class 3, and very small numbers of the other classes (Table 10.3). The most intense cushion activity, i.e. the formation and opening of flowers, occurred at the beginning of the first rains, before the period of basidiocarp production. A disease index comparing the number of diseased cushions to the number of active cushions for 1988 was 0.03 and for 1989 0.05, showing a very low level of cushion infection in both years. However, the incidence of diseased cushions did increase from 1988 to 1989.

TABLE 10.3
Annual totals for types of infection occuring on flower cushions at El Piñal

Infection class[1]	1987	1988	1989	Mean
1	25	31	17	24.3
1 & 2	2	0	0	0.6
1 & 3	2	3	9	4.7
1 & 2 & 3	5	6	0	3.7
2	6	11	0	5.7
2 & 3	2	0	6	2.7
3	19	15	16	16.6

[1] 1 single flower brooms; 2 chirimoya fruit; 3 vegetative cushion brooms.

Fruiting and Fruit Infection

Fruit set occurred between weeks 17 and 23 of 1988 and between weeks 19 and 29 of 1989. Approximately 20 weeks are required for fruit to develop and mature, so the greatest number of mature fruits were harvested between weeks 34 and 46 in both years studied. A total of 1,544 fruits was produced in the main harvest period in 1988, 80% of which were diseased with witches' broom. In the following year, when 1,296 fruits were produced, losses to witches' broom were 96%. Analysing the total production over the whole year (Table 10.4), losses to *C. perniciosa* were similar to those during the main harvest. The high losses recorded at fruit level are because the fruits are produced during periods when basidiocarps are present and are therefore exposed to inoculum for long periods.

TABLE 10.4
Relationship between fruits produced and losses caused by witches' broom at El Piñal

Year	Mean number of pods/tree			% diseased pods witches' broom
	Total	Healthy	Witches' broom	
1987	28.8	17.8	11.0	38.2
1988	72.1	27.8	44.4	61.5
1989	61.7	11.4	50.4	81.6

CONCLUSIONS

1. Rain is the determining climatic factor for the formation of basidiocarps, and most were produced during the second half of each year.

2. Many basidiocarps were produced on vegetative brooms suspended in the upper and middle canopy, and very few at soil level (almost nil). Witches' broom infected pods suspended and placed on the soil did not produce basidiocarps.

3. Most shoot and flower cushion activity escaped attack by the pathogen, because growth occurred before the period of basidiocarp production. There was some overall similarity between the two years in flushing and flowering patterns.

4. The number of vegetative infections increased greatly through the three years of the experiment, in 1989 mainly in relation to the increase in vegetative activity in comparison to 1988. The number of cushion infections did not increase by a significant amount.

5. Fruits were exposed to inoculum right from the cherelle stage and many became damaged by witches' broom as a result, which severely reduced yield.

6. The experimental area which previously had been managed under a regime involving broom removal, after the three years of the study had become unproductive because of the

damage caused by the pathogen, with fruit losses over 80%, and an mean of 234.4 and 68.4 vegetative and cushion brooms per tree in the experimental area.

MANAGEMENT IMPLICATIONS

1. Brooms should be removed before the rainy season during March and April when shoots showing signs of witches' broom infections can be taken off at the same time as old brooms which would begin to sporulate in the succeeding season. This practice is a very effective measure for significantly reducing basidiocarp production during the period of fruit formation.

2. It is not worth collecting and burning brooms and *C. perniciosa* infected fruits outside the plantation at the time of their removal. At soil level this material does not constitute a source of inoculum as it does not produce basidiocarps, and decomposes rapidly because of soil wetness and saprophyte activity.

3. The use of chemicals should be specifically directed to the protection of fruits, and confined to the period of fruit formation.

4. Combination of the practices of removing sources of basidiocarps and protecting the fruits with fungicides would significantly diminish damage to the fruits.

Chapter 11

COMPARATIVE EPIDEMIOLOGY OF THE WITCHES' BROOM PATHOSYSTEM

R.A. Schmidt, S.A. Rudgard, A.C. Maddison & T. Andebrhan

INTRODUCTION

The subject of this chapter is the analysis of the combined information from all IWBP Comparative Epidemiology sites. Summarized data from the nine sites are compared to give a better understanding of the anatomy of the witches' broom epidemic. From this analysis it is possible to refine or re-examine relationships identified at individual sites (see Chapters 5–10), to examine new relationships, and to document important natural variability among sites and years; all with disease management in mind.

The nature and scope of the comparative epidemiology experiment were explained in Chapter 3, together with the restrictions which this complex perennial crop pathosystem placed on the study. For example, incubation, latent and infectious periods for witches' broom, unlike many diseases, are variable and estimates of them are imprecise. Since only two years' data were available it would not be meaningful to compare disease progress curves and infection rates (Vanderplank, 1963) among locations as in Kranz's (1974, 1978) model for comparative studies. The IWBP comparative analysis is often similar to that of Cox & Large's (1960) summary of potato blight epidemics. The primary objective was to identify important epidemiological parameters and processes (Palti & Kranz, 1980) and to compare these critical features among sites. As appropriate to the data, causal relationships were sometimes examined.

Standard methods (see Chapter 4) facilitate comparative analysis, but there is important natural and experimental variability among sites and years which temper the analysis and, more important, the conclusions. In addition to the great biological variability in this pathosystem, identified here, other important sources of variability occurred. Trees in the IWBP study varied from 7 to 30 years of age, 3.1 to 7.5 m in mean height, 2 x 2 to 4 x 4 m spacing, and from one to several clones or hybrids. Overstorey shade ranged from 0 to 150 trees/ha. The proximity of adjacent cocoa, amount of disease in neighbouring plantations and the efficiency of initial phytosanitation (broom removal) varied among sites. Because personnel differed among sites, experimental error was present, e.g., subjective judgment was required in the selection of host material for potential inoculum sources. Such are the

Disease Management in Cocoa: Comparative epidemiology of witches' broom. Edited by S.A. Rudgard, A.C. Maddison and T. Andebrhan. Published in 1993 by Chapman & Hall, London. ISBN 0 412 58190 6

realities of large field studies based in many countries. Finally, the primary sample size (10 trees/site) was a realistic workload, but was very small relative to the true population and may not adequately represent the actual range of variation in the population.

The methods of analysis vary. In some cases, quantitative data or indices are compared in tables or graphs without statistical analyses. Other data lend themselves to statistical treatments, including multivariate analyses. In several instances results of quantitative analyses, although inconclusive, are included. Nine of ten sites were included in this analysis; the Llanos site in Colombia was not included since these data were incomplete.

This chapter considers in turn, climate, basidiocarp production, phenology of susceptible host tissue, disease dynamics, and infection efficiency and closes with discussion and conclusions and preliminary implications for disease management. Final disease management recommendations are left for Chapter 15.

CLIMATE

Long-Term Climate

The ecosystem in which wild cocoa is found is that of a moist tropical forest, with well-distributed rain and barely perceptible seasons. Cocoa is cultivated in a range of climates, some of which differ considerably from that just described, as the climate data from the IWBP sites show.

"Climograms" derived from long-term, average monthly temperature (°C) and rain (mm) at the official meteorological station nearest to each site, provide a useful comparison of typical climates (Figure 11.1). Rain (season and amount), clearly differs between sites. With few exceptions, temperature profiles are less remarkable, i.e., temperature is less variable and does not appear to limit the epidemic.

Ouro Preto, Altamira, Tomé Açu and Pichilingue exhibit distinct alternating wet and dry seasons, each lasting approximately six months. Táchira and Trinidad also have distinct wet and dry periods but they occur 4–5 months later in the year, as these sites are further north. Manizales is distinguished by the absence of a regular dry season; and average rains exceed 100 mm each month. By contrast, Pichilingue and Ouro Preto have a well-marked dry period with only minimal amounts (<50 mm/month) of rain. In the wettest periods rain exceeds 400 mm/month in Tomé Açu, Táchira and Pichilingue. Total rain (Table 11.1) is greatest in Tomé Açu and Táchira, averaging 2,800 mm/year, and least (1,800 mm/year) in Trinidad and Grenada.

The climograms indicate a variation in average monthly temperature of approximately 22–27°C among sites, a range favourable for disease development (Chapter 2). Tomé Açu is several degrees warmer than other sites throughout the year and Manizales, the highest in elevation at 1,000 m, is several degrees cooler, with average monthly temperatures all below 24°C. The transition from wet to dry season in Pichilingue, Ouro Preto and Trinidad is marked by a decrease in temperature.

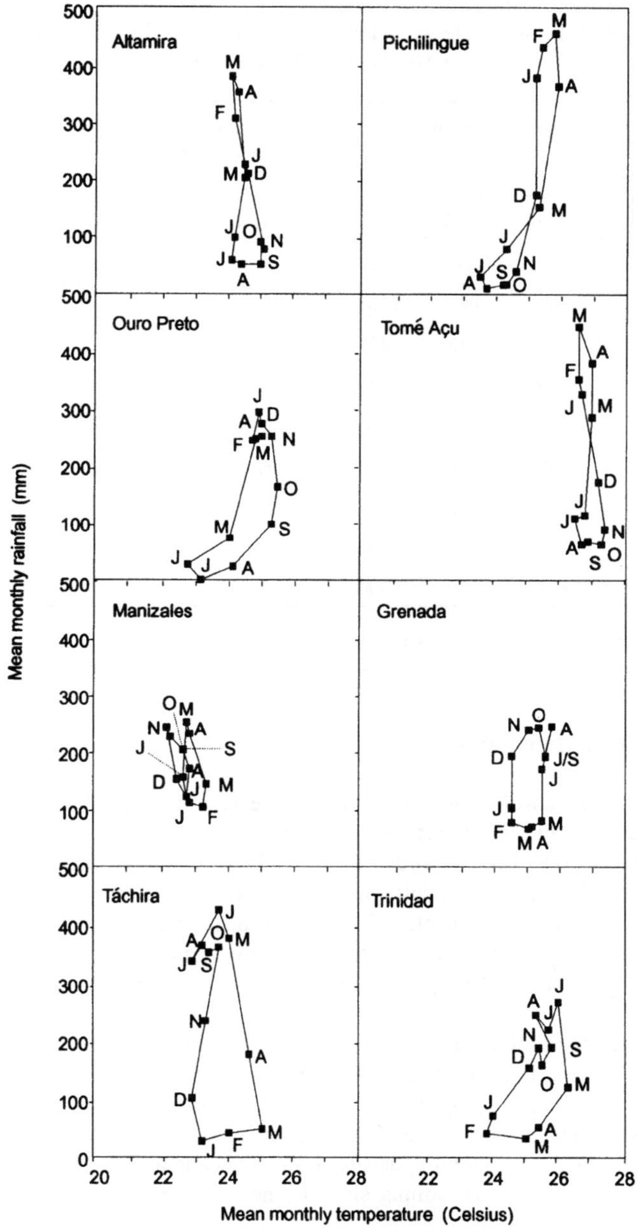

Figure 11.1. Climograms of mean total rain and average mean temperature by month from long-term weather records at the meteorological station nearest the IWBP site. Months are abbreviated as their first letter, and follow consecutively on the line beginning with January (J).

Table 11.1
Total annual rain and number of months with <100 mm and <10 mm of rain from normal climate records[1] and weather at IWBP study sites[2]

Site	Total rain (mm)			Number of months with <100 mm rain			Number of months with <10 mm rain	
	Normal	Year 1	Year 2	Normal	Year 1	Year 2	Year 1	Year 2
Tomé Açu	2,474	2,768	1,630	4	4	7	1	0
Táchira	2,800	4,002	2,167	3	4	4	2	0
Pich. Station	2,179	2,513	1,653	6	6	7	4	4
Pich. Farm	2,179	2,519	1,649	6	5	7	4	4
Manizales	2,127	1,876	1,996	0	7	4	0	0
Altamira	2,137	2,178	1,857	6	4	5	1	0
Ouro Preto	2,021	1,715	1,897	4	5	4	1	4
Trinidad	1,751	1,465	1,346	4	5	4	4	1
Grenada	1,875	1,671	1,794	4	6	4	0	0

[1] long-term records from nearest meteorological station; [2] data from IWBP sites, years 1 or 2 are not always the same calendar years among sites, and do not necessarily start in the same month.

Weather at the Experimental Sites

Comparisons of the meteorological data collected at each site during the IWBP studies show appreciable differences in rain at certain sites between the two years, and also differences in comparison with the long-term records (Table 11.1). Total rain differed greatly between years at Tomé Açu, Táchira and Pichilingue but, on the basis of the year-to-year variability seen in the long-term records (Chapters 5-10), such differences are not unexpected. Total rainfall was much less (≥ 400 mm/year) than the long-term average in five instances, and much more (nearly 1,200 mm/year) than normal at Táchira in year 1.

The number of months with less than 100 mm of rain was less variable among sites and years than total rain, but with notable exceptions (Table 11.1); drier months were more numerous than normal at Tomé Açu in year 2, and at Manizales in both years. Comparisons of the number of months with less than 10 mm of rain (Table 11.1) emphasizes the consistent dry season (4 months) at both Pichilingue sites both years and at Ouro Preto and Trinidad one year. Except for Táchira, the remaining sites did not have more than one month with less than 10 mm rain.

Temperatures at the sites (Chapters 5-10) were similar to the long-term averages, e.g. temperatures were relatively cool in Manizales and warm in Tomé Açu. There were differences between years at a given site; the minima in Manizales were higher than usual in year 2, and both Pichilingue sites were cooler than normal in the period September-December of year 1. Nevertheless, temperature differences were small and unlikely to limit disease development. While recognizing that temperature affects the rate of important epidemic processes and can profoundly influence surface moisture, which drives the infection process, nevertheless, temperature was not analysed further since it did not provide meaningful data to compare the epidemics, especially with respect to disease management.

Long–term climograms are useful to compare sites but these averages concealed considerable variability among years at the IWBP sites, particularly in the amount and distribution of rain. Greater variation among years occurs in "atypical" periods, e.g. during strong "El Niño" events, or hurricanes. Comprehensive disease management recommendations must at least consider the typical yearly variability in important climatic parameters, in this case, rain.

BASIDIOCARP PRODUCTION

Quantity of Basidiocarps

Open turgid basidiocarps were counted each week, when possible following rain, except in Manizales, where both turgid and dry basidiocarps were counted. All basidiocarp data are from a fixed number of sample units and do not reflect the actual number of such units in a tree or plantation.

Sources in Manizales produced twice the numbers of basidiocarps (1.44/object/week) that occurred on the second most productive site (Trinidad), and nearly ten–fold that of the least productive site (Pichilingue Farm, with 0.20/object/week) (Table 11.2). Irrespective of Excluding Manizales (where non–turgid basidiocarps were included), there was a four– to five–fold difference among sites in the average number of basidiocarps produced. Considerable variability in basidiocarp numbers also occurred between years at a site.

Vegetative brooms suspended in tree canopies (aerial) produced as many basidiocarps (0.29/object/week) as the other four sources combined, when averaged for all sites and years (Table 11.2). Basidiocarp production was not statistically different on the sets of brooms suspended at two different heights in the canopy, and these data were averaged into one estimate for aerial vegetative brooms. Attached cushion brooms (aerial) were the second most prolific source of basidiocarps (0.09/object/week), and production on this source was comparatively high at the two Pichilingue sites and at Manizales. Pods were the least productive source at all sites, but appreciable numbers of basidiocarps occurred on suspended pods (aerial) at Manizales, Altamira, Tomé Açu and Táchira.

In addition to being the most productive source overall, suspended vegetative brooms were also the most consistent sources among sites. Basidiocarp production from other sources, particularly vegetative brooms and pods on the ground, varied greatly among sites and sets (Table 11.2). At Manizales, where ground sources were uncovered, i.e., where natural fallen litter was removed, basidiocarps were abundant. At other sites litter, and in some cases ground vegetation, covered brooms on the ground. These covered brooms produced few basidiocarps in their first year of exposure in 10 of 16 site–year combinations. Possible reasons for the marked difference in behaviour from successive first–year sets include climate, the density of tree canopies in the plot (i.e. microclimate), and the timing of placement of the sets in relation to major leaf-fall events. By the second year, with the exception of Manizales, most of the ground sources had decayed and the only instances of basidiocarp production on the ground were for brooms at Altamira and Táchira.

TABLE 11.2
Average number of open basidiocarps of *C. perniciosa* produced on potential inoculum sources of cocoa (basidiocarps/object/week X 100)

Site	Year	Set[1]	Brooms			Diseased pods		All sources	Mean
			Vegetative		Cushion				
			Aerial[2]	Ground	Aerial	Aerial	Ground		
Manizales[3]	1	1	52	34[5]	30	9	43[5]	168	
	2	1	33	13[5]	16	12	23[5]	97	
	2	2	46	29[5]	46	12	35[5]	168	144
Trinidad	1	1	110	0	10	5	0	125	
	2	1	47	0	24	0	0	71	
	2	2	25	0	0	8	0	33	76
Altamira	1	1	36	11	3	7	3	60	
	2	1	20	3	2	6	0	31	
	2	2	34	21	3	10	6	74	55
Pich. Station	1	1	18	0	8	5	2	33	
	2	1	3	0	4	0	0	7	
	2	2	32	37	15	9	5	98	46
Tomé Açu	1	1	37	10	7	11	0	65	
	2	1	39	0	0	16	0	55	
	2	2	11	0	0	0	0	11	44
Táchira	1	1	22	0	ND	6	6	34	
	2	1	49	7	ND	5	0	61	
	2	2	23	3	ND	8	0	34	43
Grenada	1	1	35	1	3	ND	0	39	
	2	1	18	3	3	ND	7	31	35
Ouro Preto	1	1	13[4]	0[4]	5[4]	0	0	18	
	2	1	12	0	3	5	0	20	
	2	2	16[4]	23[4]	4[4]	5	0	48	29
Pich. Farm	1	1	9	3	15	2	0	29	
	2	1	10	2	0	0	3	15	
	2	2	6	0	11	0	0	17	20
Mean			29	8	9	6	5	55	

[1] data were recorded on object set 1 for two years and on newly selected objects in the second year (set 2) for one year. All numbers rounded to nearest whole number; [2] sources were cut and suspended in the tree canopy (aerial) or placed on the ground; [3] only open (turgid) basidiocarps were counted except in Manizales where all basidiocarps were counted; [4] brooms suspended while green; [5] ground brooms and pods not covered by litter; ND is no data available.

In contrast to sources on the ground, aerial sources remained productive in their second year in the field. Usually fewer basidiocarps were produced in the second year,

Station, brooms exposed for a third year produced very few basidiocarps. Cushion brooms were relatively prolific at Manizales, Trinidad and both Pichilingue sites. Appreciable numbers of basidiocarps formed on suspended pods at Manizales, Altamira, Tomé Açu and Táchira.

Duration of Basidiocarp Production

The total number of weeks in a year when basidiocarps were produced on any source varied from 50 weeks in Manizales, to 8 and 9 weeks in Ouro Preto and Tomé Açu, respectively (Table 11.3). This production did not necessarily occur in successive weeks, but generally (excepting Manizales) basidiocarps occurred continuously during a principal period of production. In only 5 of the 18 site–year combinations was the duration appreciably less than five months. Within this principal period of production, duration of production on specific sources varied greatly among sites and broom sets, but suspended vegetative brooms produced basidiocarps in more than 70% of the weeks. Cushion brooms produced persistently, except at Grenada and Tomé Açu (Table 11.3). The period of activity for suspended pods was usually shorter and more variable than that for the other aerial sources, however, this period was relatively long (≥ 6 months) at Manizales and Altamira.

Pods and vegetative brooms on the ground at Manizales were active for prolonged periods in both years, as were brooms placed on the ground at Altamira. At Pichilingue Station and Grenada, persistent sporophore production occurred on ground brooms in only one year, and at other sites production on ground sources was short–lived, or absent (Table 11.3). With a few exceptions, the period of basidiocarp production on suspended brooms encompassed that on ground sources.

Basidiocarp Production and Broom Age

After resumption of the rains at sites with distinct wet and dry seasons, basidiocarps appeared later (17 week) on suspended vegetative brooms formed in the previous year ("first year brooms"), than on older brooms (9 week) which had formed two years ago and been exposed in the field throughout the previous year ("second year brooms") (Figure 11.2). At certain sites, basidiocarps appeared on second year brooms within a week of the first appreciable rain. More brooms were active and more basidiocarps were produced from first year brooms, but the duration of basidiocarp production was not statistically different between first and second year brooms.

Green brooms do not produce basidiocarps (Chapter 2). Data from vegetative brooms cut and suspended while green in Ouro Preto indicate that in comparison with necrotic brooms, "immature brooms" required longer (18–26 week) to produce basidiocarps, and produced fewer basidiocarps on fewer brooms. Only 35% of these "immature brooms" ever produced basidiocarps.

Effects of Climate on Basidiocarp Production

The need for a "maturation" period prevented basidiocarp production in first year brooms early in the wet season when weather conditions were favourable. Also, second year brooms, especially those on the ground, lost their ability to produce basidiocarps due to decomposition

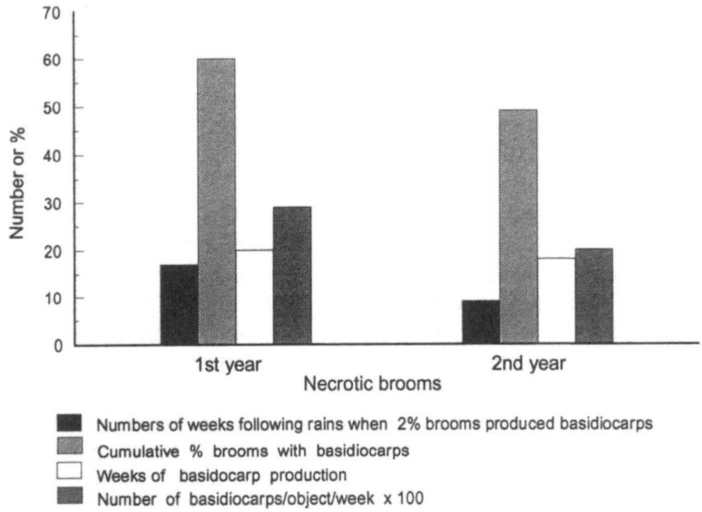

Figure 11.2. Basidiocarp production by *C. perniciosa* on first and second year necrotic brooms of cocoa suspended in the canopy (mean for all sites).

as the wet season advanced, and failed to produce basidiocarps later in the season even though climatic conditions were suitable. Such complications make it more difficult to clarify the relationship between climate and basidiocarp production. The complexity of the relationship between macroclimate and the water relations of diseased tissues of varying types and sizes, positioned in a multitude of locations within the plantation gives further difficultires in interpretation. For example, intermittent abundant heavy rain may not inhibit basidiocarp production on brooms exposed in the upper canopy because these brooms dry periodically, but brooms on the ground might remain waterlogged and unable to produce basidiocarps. Conversely, after isolated rains, brooms on the ground may be wet enough to produce basidiocarps, while canopy brooms dry quickly and do not produce basidiocarps. Rocha & Wheeler (1982, 1985) have discussed factors, including water balance, affecting basidiocarp production by *Crinipellis perniciosa*.

It is not surprising that few significant relationships were reported in the individual site analyses (Chapters 5–10) between weekly variations in the total numbers of basidiocarps and the amount of rain. One weak but significant relationship (R^2 = 0.35) occurred at Manizales where almost year–round basidiocarp production was associated with evenly distributed, though relatively light rains. At the other sites, where wet and dry seasons were well defined, basidiocarps were more numerous in the second half of the rain period in most years, and few or no basidiocarps appeared during the dry season. Basidiocarps usually were not evident in months with less than 50 mm of rain, exceptions occurred at Pichilingue and Táchira, where isolated rains were sufficient to stimulate basidiocarp production.

TABLE 11.3
Duration in weeks of basidiocarp production by *C. perniciosa* on all sources on cocoa and percentage duration on specific sources in their first year of exposure[1]

Site	Set	Duration[2]	% of total duration				
			Vegetative brooms		Attached cushion brooms	Pods	
			Aerial	Ground		Aerial	Ground
Manizales	1	48	98	88	100	67	77
	2	50	86	78	98	62	74
Grenada	1	25	100	0	4	ND	4
	2	48	100	29	23	ND	6
Altamira	1	25	92	76	56	48	24
	2	27	85	70	74	48	33
Pich. Farm	1	27	81	19	89	22	0
	2	15	73	0	87	0	0
Pich. Station	1	24	92	0	83	21	8
	2	17	82	47	82	53	6
Táchira	1	27	100	0	ND	22	4
	2	18	94	5	ND	11	0
Trinidad	1	20	85	0	100	5	0
	2	19	100	0	ND	11	0
Tomé Açu	1	23	87	22	26	43	0
	2	8	75	0	25	0	0
Ouro Preto	1	9	100	0	67	0	0
	2	17	88	18	53	12	0

[1] sets 1 and 2 were exposed in successive years; [2] on all sources; ND is no data available.

Analysis of basidiocarps production in dry periods (Table 11.4) shows that basidiocarps were observed in fewer than 20% of weeks with <10 mm of rain, and in only 6% of instances (mainly at Pichilingue and Táchira) where two weeks with <10 mm occurred successively. Basidiocarp production in dry periods may be stimulated by high humidity or abundant dew on brooms exposed to the sky or by light showers or mists not recorded on the rain gauge.

The role of climate in basidiocarp production was examined more closely with multiple regression analysis during a 12 week period of abundant sporulation at each site. The relationships among sites between the mean basidiocarp index (basidiocarps/broom/week) and nine meteorological variables were investigated. No significant regressions were found for pods in the canopy or brooms on the ground. However, the *F* ratios with two independent variables were highly significant ($P < 0.01$) for suspended vegetative brooms ($R^2 = 0.73$), cushion brooms ($R^2 = 0.74$), and pods on the ground ($R^2 = 0.85$). One independent variable, "mean number of hours with relative humidity <70%" appeared, with a positive sign, in all three equations. For aerial cushions and brooms and pods on the ground, the second variable was "percentage of rain starts between 00.00 and 06.00h", while for suspended brooms it was "percentage of rain starts between 06.00 and 12.00h", both with positive signs. These analyses suggest that greater basidiocarp production was favoured by drier days in the rainy season, and by early morning rain.

Rain (amount, frequency and duration) is the primary climatic parameter influencing basidiocarp production. However, the interaction of macroclimate with cocoa trees results in a diverse crop climate and diverse moisture conditions on potential inoculum sources which also vary in age and location in the plantation. As a result, it was not possible - and is perhaps unrealistic - to develop a definitive model to predict the timing and quantity of basidiocarps in relation to rain.

TABLE 11.4
Number of weeks within a two-year period when basidiocarps of *C. perniciosa* were present on sample sources in cocoa plantations during periods with little or no rain

Sites	Weeks without rain			Weeks with rain ≤10mm		
	Total	Basidiocarps present		Total	Basidiocarps present	
		Total	≤10 mm rain previous week		Total	≤10 mm rain previous week
Turgid basidiocarps						
Tomé Açu	30	6	1	14	1	0
Altamira[1]	8	1	0	14	3	1
Ouro Preto	29	1	1	16	5	0
Trinidad	24	0	0	18	5	0
Pich. Station	25	4	3	32	5	2
Pich. Farm	22	1	0	36	6	3
Táchira	28	4	2	14	7	2
Grenada	5	3	0	20	13	5
All basidiocarps						
Manizales	13	12	7	20	20	10

[1] 98 weeks only were sampled.

PHENOLOGY OF SUSCEPTIBLE HOST TISSUE

It is recognized that the phenology of growth in cocoa is a critical factor in the development of witches' broom (Baker & Crowdy, 1943). *C. perniciosa* only infects young growing tissues – shoots, flowers or pods. Research has provided information on phenology related to growth and yield (Alvim, 1977) and on the effects of shading on witches' broom disease (Andebrhan, 1985*b*). Phenology is also discussed in previous chapters of this monograph. This section summarizes the periodicity and quantity of susceptible host tissues.

Vegetative Shoot Growth

Shoot growth occurred in distinct flushes (Chapters 5–10). Flush frequency varied from three per year in Grenada, to six or seven per year in Altamira. Bud activity in the periods between these principal flushes also varied among sites. In Altamira susceptible vegetative tissue was present year–long, while at Manizales and Grenada there was prolonged inactivity, having an average of 12–13 biweekly periods per year without appreciable flushing. At most sites short but intense flushes, signifying synchrony among the sample branch units and trees,were interspersed with less intense flushes on a few trees or parts of trees. Flushing was particularly well–synchronized on the single clone at Manizales.

TABLE 11.5
Shoot activity index[1] and flower cushion activity[2] of cocoa in relation to tree age, and shade tree planting densities

Site	Cocoa tree age (yr)	Shade trees /ha	Shoot activity index		Flower cushion activity			
					Sum active (x1000)		Maximum no. active (x1000)	
			Yr 1	Yr 2	Yr 1	Yr 2	Yr 1	Yr 2
Tomé Açu	7	110	2.4	0.8	22.4	17.9	1.6	2.1
Altamira	9	0	4.9	3.7	44.4	42.0	2.4	2.5
Ouro Preto	13	50	2.7	1.9	54.1	50.5	2.6	2.7
Trinidad	13	30	2.6	1.6	19.4	11.8	1.5	1.1
Pich. Farm	16	60	2.8	2.7	6.7	4.7	0.7	0.7
Manizales	18	40	1.9	1.2	6.5	3.5	0.4	0.3
Pich. Station	19	30	2.2	1.9	4.4	2.8	0.5	0.4
Táchira	25	150	0.9	1.0	2.2	1.3	0.6	0.1
Grenada	30	10	*	*	5.0	2.4	0.3	0.2

[1] sum of flush points/sum of terminal buds; [2] cushions scored as active with flower buds and/or open flowers; * incomplete data for annual index.

At Pichilingue, where the Station and Farm sites were separated by only 5 km and experienced the same climate, the timing of flushes in the two sites was similar, though coincident flushes differed in intensity. For a given site, the number, timing and intensity of major flushes often differed between years.

The quantity of active flush points was greatest in the young and unshaded trees at the Altamira site, where flushing was also most frequent. Older plantations and those with heavy shade flushed less. The trees at Táchira flushed least, giving rise to only 10% of the number of flush points as at Altamira.

Numbers of potential flush points within the sample units changed with time as a result of growth and infection, hence a shoot activity index (cumulative total of flushing buds/number of terminal buds) was calculated to compare among years and sites (Table 11.5). At sites where initial pruning occurred, the activity index was generally higher in the first year. Pruning stimulated bud burst in the first year, and its absence before the second year allowed subsequent self-shading, perhaps limiting flushing. At Pichilingue, where there was no initial pruning, the activity index was nevertheless reduced in the second year at both sites, perhaps as a result of climate differences. Vegetative flushing analysed with time series ARIMA models showed a lack of regular periodicity even at Pichilingue where there was no maintenance pruning to influence natural patterns.

Flushing in year 1 (expressed as the logarithm of cumulative active flush points) and tree age were negatively correlated ($P < 0.01$, F ratio of 23.9, $R^2 = 0.77$) among sites. When shade tree density (trees/ha) was included the F ratio was 24.9 ($R^2 = 0.89$) and the coefficient was negative. Flushing was not significantly related to tree age or density in the second year. The logarithm of the shoot activity index was similarly related to tree age and tree density in the first year, but not in the second year.

Flowering

The presence of flower buds and open flowers was registered as cushion activity (Table 11.5). There was one principal period of activity each year, except at Altamira and Manizales where there were two. Brief lapses in cushion activity were common, but some flowers were present year-long.

The maximum number of cushions active at one time on the 50 m of marked branches and trunk varied between 2,715 at Ouro Preto and 149 at Táchira. Both maximum and cumulative cushion activity were negatively related to tree age (Table 11.5) and, as with flushing, dense shade apparently suppressed cushion activity even in the young trees at Tomé Açu. Among all sites, the logarithms of cumulative active cushions was significantly related to tree age; $F = 11.7^*$ ($R^2 = 0.63$) and $F = 14.1^*$ ($R^2 = 0.67$) for years 1 and 2, respectively. Inclusion of shade tree densities accounted for slightly more variation; with $F = 8.1^*$ ($R^2 = 0.74$) and $F = 8.4^*$ ($R^2 = 0.74$) for years 1 and 2, respectively.

Fruit Production

Maximum pod setting did not always occur when cushion activity was greatest, consequently pod production was not closely associated with the amount of flowering. As judged by pod harvest data, there were one or two principal periods of pod production each year; Manizales, Pichilingue Station and Táchira had two such periods, while Grenada, Pichilingue Farm and

Ouro Preto had one principal period each year. Altamira and Tomé Açu had one or two periods depending on the year. The pattern in Manizales may have been modified by the removal of all developing pods initially. Trees at Altamira produced pods throughout the year, whereas in Trinidad, Grenada and Pichilingue Station there were periods when few or no pods were produced. An average of 45–50 pods, healthy and diseased, were harvested from each tree, varying from 100 in Ouro Preto to 18 in Trinidad. Productivity was not related significantly to tree age, shade tree density or rainfall. The mean number of pods increased at some sites, but decreased at other sites in the second year.

DISEASE DYNAMICS

Witches' broom results from the coincidence, under favourable weather conditions, of susceptible cocoa tissues with basidiospore–producing basidiocarps of *C. perniciosa*. So far, the comparative analysis has considered these elements of the disease triangle separately, but it is their interaction which determines disease incidence and severity. An understanding of the qualitative and quantitative nature of their coincidence is a prerequisite to predicting the time of occurrence and magnitude of disease, and to developing disease management strategies.

To examine the interplay among the parameters in the disease triangle, it is necessary to know when infection occurs, not when disease symptoms appear. In many diseases the incubation period, i.e., time from penetration to symptom expression, is consistent and precisely known for a range of conditions. The studies on witches' broom in mature trees (Chapter 2) show that the incubation period varies widely according to the type and age of tissue infected, inoculum dose and environment, but precise estimates are lacking for many situations. Moreover, while the incubation period in pods at a given site appears reasonably constant, simultaneous infection on shoots and cushions is known to produce symptoms whose appearance spans a period of weeks or even months. Therefore, our field data cannot identify the precise week of infection nor, parenthetically, the amount of inoculum and susceptible host tissue which occurred at this unspecified time. Nevertheless, some analysis (Table 11.6) was attempted for each of the three infection courts, using a range of reasonable estimates of the incubation period to allow for the "lag" in the appearance of disease symptoms. It was generally assumed that climate favourable for basidiocarp production was also favorable for infection.

Regression analysis was used to examine relationships between the bi–weekly records for numbers of symptoms and basidiocarp numbers lagged by 6 or 8 week for shoot and cushion disease, and by 10, 12 or 14 week for pod disease. Generally only a small part of the variation in disease incidence was explained, and the only significant regressions for vegetative and cushion symptoms ($R^2 = 0.46$ and 0.34, respectively) were obtained from data at Altamira with an incubation period of 6 week (Table 11.6). Significant regressions for pod symptoms were found only with data from Ouro Preto and Trinidad. At Manizales, where it was cooler than the other sites, longer incubation periods yielded better relationships among the variables when compared with shorter incubation periods.

Regressions of amounts of disease using two independent variables (basidiocarp numbers and amount of susceptible tissue) were only appreciably better than using one variable only (basidiocarp numbers) for variations in pod disease. An R^2 of more than 0.80 was found at Ouro Preto and Táchira, and the only regression for pods that was not significant was that for Manizales (Table 11.6). Although quantitative relationships were poor at Manizales, there was good coincidence in the timing of infections on the three infection courts when the incidence of shoot and cushion symptoms were lagged 8 week and the percentage of infected pods 14 week (Figure 11.3).

TABLE 11.6
Relationships among the quantity of diseased cocoa tissue and basidiocarps of *C. perniciosa* and susceptible host tissue as determined by the coefficient of variation (R^2) from single and multiple regression analyses

Site	Vegetative disease[1]		Flower cushion disease[2]		Fruit disease[3]	
	b'carps	b'carps and active points	b'carps	b'carps and active cushions	b'carps	b'carps and young fruit
Altamira	<u>.46</u>	.47	<u>.34</u>	.38	.05(10)	<u>.62</u>(10)
Tomé Açu	.15	.26	.21(8)	.24(8)	.15(10)	.39(10)
Ouro Preto	.31	.34	.06	.17	<u>.49</u>	<u>.87</u>
Trinidad	.01	.01	.02	.47	<u>.37</u>	<u>.77</u>
Táchira	.06	.06	.03	.03	.01	<u>.81</u>
Pich. Farm	.36	.37	.21	.39	.20	<u>.65</u>
Pich. Station	.15	.23	.05	.07	.04	<u>.53</u>
Manizales	.01(8)	.03(8)	.04(8)	.06(8)	.24(14)	.31(14)

Independent variables (basidiocarps and susceptible tissues) were lagged in time (week) by the assumed incubation period for each tissue, which was [1] 6 week, [2] 6 week and [3] 12 week except where given in parentheses; R^2 values significant at 0.01 level of probability are underlined, others are non-significant.

The lack of good quantitative relationships in these data may have explanations in addition to variability in incubation period. In theory, the relationship between numbers of basidiocarps and amount of disease may not be linear – possibly because in many situations even relatively few basidiocarps provide abundant basidiospores. A further uncertainty in the analysis was how well basidiocarp production on our relatively small samples of diseased tissues reflected that on natural sources in the vicinity. Basidiocarps on the surrounding trees in the plantation could influence the amount of disease in the sample trees (Chapter 13).

To examine the coincidence of basidiocarp production on the sample sources with disease incidence, graphs of basidiocarp production from the sets were superimposed with the appropriate lag period on graphs of pod disease. Manizales was excluded from this exercise owing to its year–round production. Pichilingue Farm was the only site where pod infection appeared to coincide with basidiocarp production. At all other sites, pod infection preceded basidiocarp appearance. This implied that early basidiocarp production, which occurred predominantly on old brooms, was not generally represented in the sample set of sources in year 1, and the broom samples taken from those sources cannot have contained sufficient older brooms, except at Pichilingue Farm. The coincidence at the end of the basidiocarp

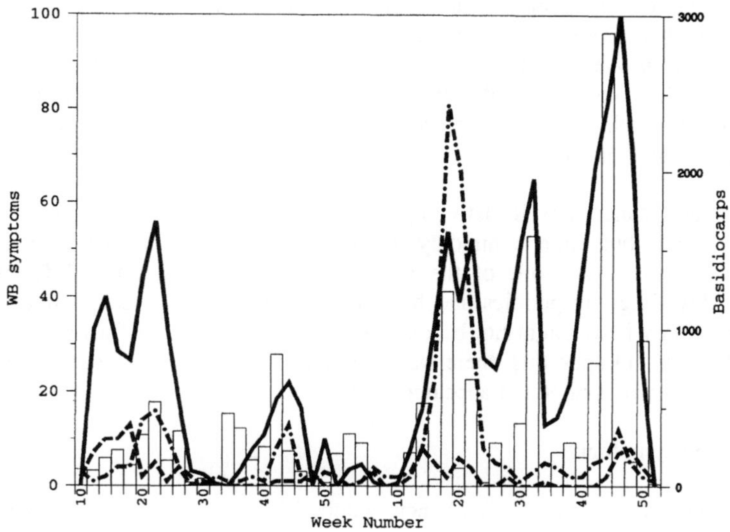

Figure 11.3. Basidiocarp production by *C. perniciosa* and witches' broom infection on cocoa during a two-year period at Manizales. Basidiocarp totals, bars; shoot symptoms 8 wk later, broken line; flower cushion symptoms 8 wk later, dotted broken line; % pods with witches' broom 14 wk later, solid line.

season was generally good, except at Táchira and Ouro Preto, where pod symptoms continued to appear when basidiocarps were absent from the samples. Pod infection during the dry season at Pichilingue in year 2 was apparently related to sporadic basidiocarp production prompted by small rain events and the drifting fogs or "garúas" (Chapter 7).

These field data are not well-suited to predict the amount of disease which will result from known amounts of susceptible host tissue and inoculum given favourable climatic conditions, but they do provide excellent information on the timing of the critical events most useful for disease management. Timing of symptom appearance is discussed further in the following section.

Symptom Occurrence in Relation to Phenology

Shoot symptoms

At most sites there was considerable variation between years in the number of vegetative symptoms recorded. Differences in amounts of inoculum and in the suitability of climatic conditions from year to year contributed to this variability, but phenology also was involved.

The marked periodicity seen in shoot growth, superimposed on large fluctuations in basidiocarp numbers from week to week, meant that a principal flush might coincide with a basidiocarp peak in one year, but not in another. At Tomé Açu, for example, in 1986 one of the main flush peaks coincided with the latter part of the basidiocarp season and many infections resulted, while in 1987 the end of wet season flush was weak and bore only half as many infections (Chapter 5).

In some instances a well-defined peak in symptom occurrence could be related to an earlier flush peak, but more commonly vegetative symptoms emerged over a protracted period, presumably as a result of the variable incubation period for differing types of symptoms. Swellings on petioles and branches are visible in 4 week, while infection of terminal or axillary buds might not become apparent until the shoot flushes again weeks or even months later. Despite this, there was a period of several months each year at all sites when few, or no, new symptoms appeared on flushes.

Flower cushion symptoms

With a few exceptions (e.g. Táchira), new witches' broom symptoms on flower cushions tended to appear over longer periods than those on shoots, as might be expected given the greater consistency of flowering relative to flushing (Chapters 5–10). But despite the more consistent activity, curves for flower cushion symptoms often did not coincide with those for basidiocarps, duly lagged, on the samples. If allowance were made for lack of early sporophore production in the first year, there was reasonable correspondence at a few sites, for example Altamira, assuming an incubation period of 6–8 week (Table 11.6, Figure 11.4). In other sites, symptoms continued to appear long after basidiocarp production on the sets ceased (Tomé Açu, Ouro Preto) or, conversely, failed to appear when sporophores and susceptible tissue had been present (Manizales, and Trinidad year 1). Prolonged incubation periods might explain the former instances, but the reasons for the latter are not clear, unless low minimum temperatures were responsible.

Compared to the situation with shoot symptoms, generally there were fewer weeks in the year when no new symptoms appeared on cushions. Nevertheless, except at Manizales, there was one period each year when there was a lull of at least 8 week in the appearance of symptoms on shoots and cushions.

Symptoms on pods

Comparisons of disease in relation to harvesting (Chapters 5–10) placed the sites in several categories. The first, where most diseased pods were harvested during the relatively concentrated main harvest period in the early dry season, included Ouro Preto, Tomé Açu and Trinidad. At these sites, although escape from disease was theoretically possible, relatively few pods were set during the dry season and less than 25% of the total healthy crop was harvested in the non-disease period, except in Tomé Açu in year 2, where the proportion was 46%.

In those remaining sites with a marked dry season, production was spread over longer periods than disease incidence, and pods maturing at the end of the dry season/beginning of the wet season escaped disease. Pods already 3 months old – when the rains began and

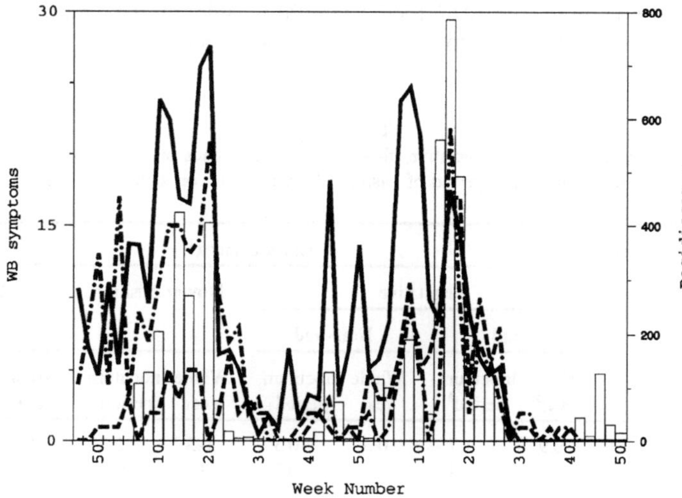

Figure 11.4. Basidiocarp production by *C. perniciosa* and witches' broom infection on cocoa during a two-year period at Altamira. Basidiocarp totals, bars; shoot symptoms 6 wk later, broken line; flower cushion symptoms 6 wk later, dotted broken line; % pods with witches' broom 10 wk later, solid line.

basidiocarps appeared – could not be damaged by subsequent *C. perniciosa* infection. At Táchira and both Pichilingue sites, 50–66% of the healthy crop was harvested in the disease-free season. Although drier periods at Manizales were less marked than at other sites, they nevertheless permitted some escape. At Altamira, the percentage of pods infected during the wet season was lower than in most other sites (see "Pod disease indices" below), and the proportion of escape pods to the total was in the range 23–29%. Pod infection by *C. perniciosa* in the Grenada site was negligible, though infrequent harvests and infection by *Phytophthora* pod rot may have led to underestimation of pod losses to witches' broom.

INFECTION EFFICIENCY

To examine disease efficiency among sites and years, disease indices were calculated (totals of symptoms/sum of active flush points, active cushions or harvested pods). Two indices were obtained for each infection court (Table 11.7), one with the denominator derived from counts during a full year, and the other from counts during the main period of basidiocarp production on the sources (the "inoculum period"). Although the two indices were highly correlated, with *r* varying from 0.86 (pods) to 0.98 (cushions), their ratio (inoculum period index/year index) ranged from 1 to more than 5 between sites.

TABLE 11.7
Witches' broom disease indices for vegetative, flower cushion and pods on cocoa for a 52-week period and for the principal period of basidiocarp production on sample sources

Site	Year	Disease index[1]							
		Vegetative				Flower cushion		Pods	
		Bud		Non-bud					
		Whole year[2]	Inoculum period[3]	Whole year	Inoculum period	Whole year	Inoculum period	Whole year	Inoculum period
Altamira	1	0.9	1.7	0.3	0.5	0.3	0.7	12	12
	2	1.8	2.4	0.5	0.7	0.2	0.3	11	14
Grenada	1	1.2	4.5	0.2	0.6	4.9	7.6	2	*
	2	5.3	7.5	0.0	0.0	18.1	18.1	8	*
Manizales	1	0.2	0.3	1.0	1.0	1.5	2.1	10	11
	2	3.1	3.6	5.0	5.9	8.3	10.1	24	41
Ouro Preto	1	1.8	3.9	0.2	0.5	0.1	0.3	29	34
	2	2.6	6.5	0.3	0.6	0.4	0.7	59	71
Pich. Farm	1	13.4	18.3	12.9	17.5	1.4	3.0	31	56
	2	2.3	3.8	6.3	10.3	1.9	3.1	28	52
Pich. Station	1	4.3	6.3	13.6	19.9	0.9	3.2	16	43
	2	3.3	4.0	7.4	8.9	2.1	4.4	20	47
Táchira	1	29.7	154.6	2.9	15.3	2.9	4.4	62	72
	2	45.7	177.2	0.5	1.8	4.0	5.7	83	*
Tomé Açu	1	3.4	9.9	1.7	4.9	0.5	0.6	30	32
	2	7.1	18.1	0.6	1.6	0.5	1.3	41	56
Trinidad	1	0.7	2.5	0.7	2.5	1.3	1.8	44	49
	2	1.3	7.3	0.0	0.0	1.8	2.2	27	35

[1] disease index = (number of diseased points/number of infection courts available) x 100; [2] disease index calculated for one year; [3] disease index calculated during the principal period of basidiocarp production; * incomplete data.

The differences in infection efficiency among sites were large and, even where there was appreciable disease every year, the index for a particular infection court at one site could vary several-fold between years (e.g. shoots, Pichilingue Farm).

Vegetative Disease Index

Initial vegetative symptoms were divided into two classes; bud infections, i.e., those affecting buds (terminal and axillary brooms), and non-bud infections, i.e., those symptoms (swelling and necrosis) on petioles, pulvini and twigs. Bud infection was most efficient at Táchira in both years (Table 11.7) and higher than average at Pichilingue Farm (year 1), Tomé Açu, and Grenada (year 2). There was no correlation between non–bud indices and bud indices (r=0.025). Non–bud infection was less efficient than bud infection, except at Manizales and in the two Pichilingue sites (Table 11.7) where non–bud symptoms were frequent. Pichilingue Farm, followed by Pichilingue Station and Táchira, exhibited the greatest number of shoot symptoms: Grenada had the least shoot infection. There were significant increases from year 1 to year 2 for bud indices at four sites out of nine (Táchira, Tomé Açu, Manizales and Grenada), and for non–bud indices at Manizales only.

Cushion Disease Index

The cushion infection index varied less among sites and years than the shoot index, but ranged from 18.1 at Grenada in year 2, to 0.1 at Ouro Preto in year 1 (Table 11.7). Cushion infection was relatively efficient in the second year at Manizales and moderately so at Táchira in both years.

Although differences in the amount of active cushions could explain differences in the numbers of cushion infections in years 1 and 2 in a few instances (e.g. Trinidad), it could not in others (e.g. Altamira and Manizales). The units at Grenada contained the greatest numbers of infected cushions, followed by those at Trinidad and Manizales. Trees at Táchira and Pichilingue Station showed the lowest flowering activity and the fewest cushion symptoms during the basidiocarp period.

Infected cushions were grouped according to the initial symptoms; those where vegetative brooms were present, alone or with other classes, those with infected chirimoyas and flowers, and those with infected flowers only. The last group was the largest over all sites: 60% and 53% for year 1 and 2, respectively. Manizales was exceptional because cushions with vegetative brooms outnumbered the other groups in both years. The clone at Manizales, SC–6, is known for its abundant production of brooms on flower cushions, a characteristic seen on other genotypes elsewhere.

Pod Disease Index

An average of 30% of the total number of pods harvested at all sites were infected by *C. perniciosa*, 57% were healthy, and the rest were lost to other diseases or pests. Losses to moniliasis (*Moniliophthora roreri*) at Pichilingue Station were 20 and 41% in years 1 and 2, respectively and at Pichilingue Farm 14 and 19%. Pods diseased with moniliasis were removed weekly at Manizales; their frequency was 1 and 8% in years 1 and 2, respectively; the disease was absent from other sites. *Phytophthora* pod rot caused losses at Grenada (16 and 24%), Trinidad (11 and 16%), Tomé Açu and Pichilingue Farm (4–8%), but was unimportant at other sites. Other losses, including rodent damage at Manizales, never exceeded 8% of the total harvest.

Variation in the infection efficiency among sites and years was less for pods than for shoots and cushions (Table 11.7), perhaps because individual pods were exposed to inoculum over longer periods than individual shoots or flowers or because pods provide more favourable conditions for infection. Pod disease indices for the year were highest at Táchira and least at Altamira and Grenada. The rankings of other sites changed between years;

indices differed appreciably for the two years at Manizales, Ouro Preto, Tomé Açu and Trinidad. Trinidad was the only site where the indices fell appreciably in the second year. Overall, the indices at Manizales and Pichilingue Station were lower than that at Pichilingue Farm, which in turn was lower than those for Trinidad, Tomé Açu and Ouro Preto. When the indices for the inoculum period were compared, the Pichilingue sites moved up the ranking because of disease escape. The appreciable losses to moniliasis and *Phytophthora* pod rots at Pichilingue and Trinidad, respectively, may have distorted the infection indices.

Data from six of the sites showed that witches' broom infected pods at all heights up to 9 m. At several sites there was a tendency for percentage disease to decrease with increasing height, but the reverse was true at Trinidad where *Phytophthora* pod rot was common on pods on the lower trunk. At Pichilingue Farm, 40% of the pods were diseased with witches' broom above 7 m, compared with 70% on the lower trunk.

Combination of Indices

Single index values were calculated for bud, non-bud, flower cushion and pod disease by combining two years' inoculum period data for the ten trees at each of the nine sites. The 90 values for the four indices were used as four column vectors in a discriminant function analysis (Lachenbruch & Goldstein, 1979). The objective was to determine if the 90 trees would be discriminated into groups on the basis of similarity in their spectra of indices and, if so, how closely derived groups matched the actual site groups. Highly significant grouping was detected by the analysis, and the first two derived functions accounted for 84% of the variation (eigenvalues of 4.92 and 3.04). At each site 7–10 trees were grouped together; trees from Pichilingue Farm and Station were grouped together, and three trees at Grenada were grouped with those from Altamira (Table 11.8).

The first discriminant function was weighted for the pod and non-bud indices, and discriminated well for all sites, with Pichilingue Farm/Station at the top of the disease scale and Altamira and Grenada at the bottom. The sites with moderate values were placed in ascending order: Manizales, Trinidad, Ouro Preto and Tomé Açu. The second function, weighted for bud index, discriminated well for Táchira. The two sites with exceptionally low pod indices, Altamira and Grenada, had bud indices close to those of the other sites and were not distinguished by the second function.

Comparisons Among Indices, and With Other Variables

Generally, linear correlation coefficients were not significant between the bud, non-bud, cushion and pod indices over the nine sites in the two years. The only significant correlation was between bud and pod indices ($r = 0.76$), and even here there were two sites, Pichilingue Station and Trinidad, at which one of the indices increased appreciably from year 1 to 2, while the other decreased.

The removal of all inoculum sources at most sites at the start of the experiment should have reduced the disease indices in the first year, if the sites were not dominated by external inoculum. Thereafter, the accumulated brooms within the site should have allowed disease incidence to increase. However, comparisons of the year 1 and year 2 indices are complicated because treatment and climate are confounded in the two years, except at the

Pichilingue sites where the treatments remained the same each year. Pichilingue Farm was not cleared of brooms in either year, while Pichilingue Station was cleared before both years (Chapter 7). Even at these two sites where the treatment was constant, the four indices did not follow the same trends. At both Pichilingue sites, cushion indices increased in year 2 while bud and non-bud indices declined. The pod index decreased slightly at the Farm site, but increased at the Station site.

TABLE 11.8
Classification of the ten sample cocoa trees at each site according to discriminant function analysis of disease indices[1] of vegetative, flower cushion and pod infections

Actual group number and site	Predicted group number								
	1	2	3	4	5	6	7	8	9
1 Altamira	10	0	0	0	0	0	0	0	0
2 Ouro Preto	0	9	0	0	0	0	1	0	0
3 Tomé Açu	0	1	7	0	0	0	2	0	0
4 Manizales	1	0	0	7	0	0	2	0	0
5 Pich. Station	0	0	1	0	6	3	0	0	0
6 Pich. Farm	0	0	1	0	3	6	0	0	0
7 Trinidad	1	1	0	0	0	0	8	0	0
8 Táchira	0	1	0	0	0	0	1	8	0
9 Grenada	3	0	0	1	0	0	0	0	6

[1] disease index = (number of infections/number of infection courts available) x 100. [Data were combined from two years, and numbers of infection courts were found only from the principal basidiocarp periods.]

Of the seven sites where increases were expected, climate and external inoculum permitting, only at Manizales and Ouro Preto did all four indices increase consistently from year 1 to year 2 (Table 11.7). At other sites, except for Altamira, it was the non–bud index which inexplicably failed to increase. The reduction in pod index at Trinidad in year 2 may have been related to the increased incidence of *Phytophthora* pod rot.

The cushion disease index was positively related to tree age, and the linear regression for year 1 was highly significant ($P < 0.01$). Index values for year 2 were fitted by the same regression, except for the values at Manizales and Grenada which were appreciably above the regression line. This apparent relationship with tree age is difficult to understand, given that the index is a proportion. Accumulation of severe cushion infection could have led to the death of the floral meristems, but it is not clear why the fewer active cushions on older trees were more diseased than the plentiful cushions on younger trees.

The discriminant function analysis grouped all trees from the Pichilingue Farm and Station sites together, even though they were of different cocoa types (Chapter 7). The three geographically separated Brazilian sites (Chapter 4) were also separated by the analysis, which again grouped different hybrids together at each site. Trees of two different TSH clones in Trinidad were not separated by the analysis, and the diverse hybrids at Táchira were grouped together. All the above provides evidence that disease indices at the IWBP sites were not dominated by host genotype, but by other site related factors. Differences in host susceptibility or pathotypes (Wheeler & Mepsted, 1982) could have influenced the indices, but the IWBP did not provide any direct measure of such effects.

Since disease indices were different among sites, relationships between disease and the causal factors were sought. No significant relationship was found between the four indices and the basidiocarp production rates per object, across the nine sites in the two years. Neither did the macroclimate data for amount and timing of rain, or air temperatures (annual means, totals or duration) reveal any significant relationships with disease indices. Individual sites, (e.g. Manizales), had a unique macroclimate, but it was not possible to link disease incidence to climate parameters. Detailed studies of microclimate prior to the IWBP showed that pod disease incidence was related to numbers of surface wetness periods at critical times at Ouro Preto (Rudgard & Butler, 1987), but IWBP disease indices were not related to the numbers of similar periods taken from the Belfort string sensors.

GENERAL DISCUSSION AND CONCLUSIONS

The epidemiology of the witches' broom pathosystem was compared over two years at nine sites in six countries. Standard methods aided data collection and reduced experimental error but natural variability, a small, but manageable, sample size, and uncontrolled experimental error typical of extensive field studies resulted in large variations in epidemiological parameters among sites and between years. The specific objective was to compare the "what, when and where" of potentially important epidemiological parameters, i.e., climate, inoculum (basidiocarp) production, susceptible host tissue, and disease dynamics. The general conclusion from these data is that witches' broom epidemiology in the humid tropics, where climatic and biotic conditions are most often favourable for disease development, is complex and not easily studied nor teased apart. The epidemic ebbs and flows, but seemingly not in response to easily-defined relationships, as is often the case for pathosystems in temperate regions or on annual crops. Nonetheless, the information obtained from comparative analyses of witches' broom epidemics is useful in the search for disease management strategies.

Climate

Seasonality and amount of rain were the distinguishing climatic features among sites and between years. Temperature differences were much less remarkable and within a range deemed favourable for disease development. Temperature can control the rate of important epidemic processes, but here rain provided a key driving variable (on/off mechanism) for the epidemic. Well-defined relationships involving rain were masked by a variable crop climate, phenology and poorly defined latent and incubation periods. The great variation in rain among sites and between years means that disease management strategies must be site specific and flexible.

Basidiocarp Production

The wet season determined the period when basidiocarps were produced, which varied from almost the whole year in some instances to several months in others; 4–5 months was typical. Young brooms formed in the preceding wet season (first–year brooms) usually produced more basidiocarps than second–year brooms when suspended in the canopy. However, basidiocarps appeared sooner on older brooms after the rains started. Third–year brooms, examined in only one site, yielded few basidiocarps. Normally, no basidiocarps were produced in periods with <10 mm of rain/week, or <50 mm/month. Nevertheless, isolated rains in the dry season were sufficient to stimulate fruiting at certain sites. Basidiocarp production tended to be more intense at sites with good daily drying conditions; generally the correlation with weekly rainfall amount was poor.

Basidiocarp production on different sources in various locations within the plantation was greatest on vegetative brooms, detached and suspended in the canopy. Attached cushion brooms and suspended pods were also moderately prolific in some instances. Production from first–year tissues on the ground was erratic; at a given site basidiocarps occurred one year, but not the next. Aerial sources continued to yield basidiocarps in their second year, but tissues on the ground decayed and were unproductive, in the second year, except at Manizales where sources were not covered by leaf–litter and decomposed slowly. The most consistent production throughout the season was on suspended vegetative brooms and cushion brooms. Pods were erratic sources of basidiocarps, with few active weeks. In one or two instances only, ground sources produced when aerial tissues were inactive. It bears repeating, because of the fate of pruned brooms, that uncovered ground sources (pruned brooms and pods) were overall 25% (12.8 basidiocarps/object/week) as productive as aerial sources.

The threat posed by a source depends on its productivity and abundance in the plantation, as well as the time of production. At the start, vegetative brooms were as numerous or more numerous than cushion brooms in most sites, but at Manizales and the Pichilingue sites cushion brooms outnumbered those on shoots ten to one (Table 4.4, Chapter 4). In the absence of sanitation, those sites would have been dominated by inoculum from cushion brooms.

Susceptible Host Tissue

There was great variation in the amount of susceptible tissue produced; both the relative amounts of shoot growth and flower cushion activity varied up to ten–fold among sites, and two–fold between years. Older trees and trees in heavily shaded plantations produced fewer new shoots and flowers compared with younger and unshaded plantations.

Vegetative flushing and flowering differed markedly in their timing at a given site. Plantations exhibited 3–7 major vegetative flushes each year, but only one or two broad peaks in flowering. Some sites had new vegetative growth in each bi–weekly sample, while other sites had periods of 3 months with no major flushes. There was some flower cushion activity throughout the 2–year period at all sites. Pod production peaked once or twice a year, and was not closely associated with the period of intense flowering. Some sites had susceptible pods throughout the year, but other sites had several months with no, or very limited, pod production. There was a five–fold range in the number of pods harvested among sites.

Disease Dynamics

Most sites exhibited one principal, extended period of disease occurrence on shoots, flower cushions and pods each year; Manizales, with year–long rains, had two such periods. Diseased pods were recorded almost year–long in Manizales and Altamira but, elsewhere there was generally a period when no new symptoms were recorded on pods.

Moderately strong correlations existed at most sites between the incidence of witches' broom disease on pods and the abundance of basidiocarps and susceptible tissues 10–14 week earlier. The correlations were less good for infection on shoots and flower cushions, perhaps because the best estimate of the incubation period, 6–8 week, was appropriate for only a proportion of the symptoms; shorter and longer incubation periods probably occurred. Some pod infection data suggested that the samples used for basidiocarp counts were not the only sources of inoculum affecting the sample trees. For example, at Pichilingue Station, disease appeared sooner or later than expected on the basis of availability of basidiocarps on the sample sets.

Disease Efficiency

Disease efficiency (amount of disease/amount of susceptible tissue) often varied several–fold between years at a given site, and by an order of magnitude or more between sites. Many factors could have caused this variability, but no causal relationship could be deduced from the data, as climate, and amounts of inoculum, susceptible tissue and diseased tissue varied greatly and inconsistently among sites and between years. Differences in host genotype susceptibility and pathotype virulence may have also played a role.

Disease indices for the various infection courts did not have similar trends from one year to the next. Non–bud disease indices fell inexplicably at many sites while bud, flower cushion and pod indices increased. Disease efficiency was exceptionally high on buds at Táchira, while the two Pichilingue sites had high non–bud indices. The comparative analyses were not able to identify definitive relationships with any certainty. For example, it was not clear in which sites or years, if any, inoculum was the principal factor that limited infection. The role of inoculum will be further explored in the comparative phytosanitation studies (Chapter 13).

DISEASE MANAGEMENT IMPLICATIONS

Numerous implications for disease management arise from the comparative analyses. Management recommendations occur in Chapters 5–10 and a comprehensive treatment of disease management with specific methods can be found in Chapters 15 and 16. Disease management is only generally discussed here under the topic of climate, basidiocarp production, host susceptibility and disease dynamics. As previously mentioned, the witches' broom epidemic in the humid tropics, where climate, inoculum periods and host phenology favour disease development – although often not in well–defined relationships – exhibits great variability among sites and years. This variability suggests that local conditions will determine the most effective management strategy. Disease management will not be necessary in some instances where pod infection is infrequent, while in other areas where disease pressure is high, disease control methods must be vigorously applied as appropriate to the economic situation.

Climate

Rain clearly drives the epidemic as it favours basidiocarp formation and host phenology, and the length of the wet season determines the duration of the epidemic. A prolonged wet season gives many opportunities for inoculum and susceptible tissues to coincide. With rain distributed throughout the year, two disease cycles per year are possible; this must be considered, when designing management strategies, particularly in relation to phytosanitation and pod loss.

Regions with a marked dry season escape infection when there is insufficient rain for basidiocarp formation. Disease escape during this period can reduce infection of cushions and pods and permit the growth of healthy foliage. The extent of disease escape will be limited if there is little host growth in the dry season, and losses may be severe because of the coincidence of concentrated activity in host and pathogen in the wet season. Practices (e.g. pruning and irrigation) which stimulate flushing, flowering and pod set during the dry season, and the selection of cocoa types with dry season activity might be investigated.

Basidiocarp Production

C. perniciosa is capable of producing large quantities of basidiocarps for long periods of time on a variety of necrotic diseased tissues in the cocoa canopy and on the ground. This greatly complicates control measures aimed at inoculum reduction. The logistics of effective phytosanitation or effective coverage by antisporulants are difficult, especially on large trees where small shoot and cushion brooms are difficult to treat. Furthermore, any pruned material left uncovered on the ground may produce a significant number of basidiocarps. Added to these complications is the potential for long–distance transport of inoculum (Chapter 13).

Host Phenology and Susceptibility

Susceptible host tissue is varied (shoots, cushions and pods), abundant and frequently available. Given the long period of basidiocarp production, it is difficult to protect potential infection courts, especially on large or rapidly growing trees. The dilution, erosion and leaching of protectant chemicals would be a further complication. However, in some instances pods offer opportunities for protection by chemicals. But even here, pods in the tops of tall trees are difficult to protect and might require protection for several months.

Because individual trees differed markedly and consistently in their amount of disease, there may be opportunities to rogue plantations or select planting material to reduce inoculum, escape disease or at least escape pod infection. Discrimination against trees which produced abundant vegetative or cushion brooms, but few pods, could aid phytosanitation.

Given the apparent difficulties of attempting to control witches' broom disease through sanitation, chemical and biological means, the development of durable genetic disease resistance would be helpful.

Chapter 12

DISEASE GRADIENTS OF *CRINIPELLIS PERNICIOSA* ON COCOA SEEDLINGS

T. Andebrhan, A.C. Maddison, R. Arias & L.A. Maffia

INTRODUCTION

Plant disease epidemics can be seen as the growth of disease in time and space. In the last three decades, the temporal aspects of disease increase and progress have been researched intensively, while spatial aspects have received somewhat less attention. Nevertheless, following the first comprehensive review of the subject by Gregory (1968), the study of gradients of dispersal and disease has advanced considerably, and knowledge about gradients has become an important requirement when defining disease management strategies.

Despite their importance to management, there are few studies reporting disease gradients for witches' broom on cocoa. An early article by Baker *et al.* (1941) in Trinidad indicated a gradient of vegetative disease in the canopy of young cocoa adjacent to a mature field, with spread up to 100 m. Evans & Solorzano (1982) obtained circumstantial evidence from seedling exposure in Ecuador indicating that viable inoculum travels at least up to 800 m, while Aragundi *et al.* (1988) followed gradients of symptoms on seedlings in an open area up to the greatest distance tested of 285 m from the source.

The studies reported below looked at disease gradients on cocoa seedlings. They complement the work on disease gradients in mature cocoa (Chapter 13). The first of the two studies was carried out in a cleared corridor in dense primary forest near Manaus, Brazil. It examined the effect of source height on the gradients obtained in seedlings, likewise placed at various heights up to 3 m. The second study (Arias, 1990) used seedlings on the ground in a small, moderately isolated cocoa plantation to detect gradients in disease at increasing distance from part of the plantation which had not received sanitary pruning to remove witches' brooms.

MATERIALS AND METHODS

Brazil

The experiment was conducted at CEPLAC's Rio Negro experimental station situated 65 km

Disease Management in Cocoa: Comparative epidemiology of witches' broom. Edited by S.A. Rudgard, A.C. Maddison and T. Andebrhan. Published in 1993 by Chapman & Hall, London. ISBN 0 412 58190 6

from Manaus, in a cleared corridor, 8 x 400 m, in dense primary forest. Wooden scaffolds, 4 m high, were positioned at distances of 1, 10, 20, 40, 80, 160 and 320 m horizontally from the inoculum source. Brooms that were ready to produce basidiocarps were suspended on nylon string and placed at the source at different heights; simultaneously in all four positions (exposure 29 March – 12 April, 1988), at 3 m above the ground (13–26 April), at 2 m (27 April – 10 May), at 1 m (16–27 May), or on the ground (7–12 July). A total of 4,000 brooms was used in each of the five experiments so that there were 1,000 brooms at each level in the "all heights" test . Basidiocarps appearing on the brooms were counted, and also the number of days when they were present.

Seedlings of the cocoa type Catongo were used because of its high susceptibility to witches' broom disease. Seeds were sown in soil–filled plastic bags in a shade house and at 6–7 weeks the growth above the cotyledon node was cut off to stimulate two growing points per plant. Four to six weeks later, seedlings with active buds were taken to the field and positioned on the ground, and at heights of 1, 2 and 3 m. For each combination of horizontal distance and height, 25 active buds were used. The seedlings and brooms in the source were shaded by a wire mesh allowing 50% light transmission, mounted on the top of the scaffolds. After being exposed in the field, the seedlings were returned to the shade house. Disease assessments were begun 2–3 weeks later by recording the presence of terminal brooms, or swellings on the pulvinus or petiole. Any seedlings that showed symptoms during the exposure period or within 1 week of removal from the field were eliminated from the trial.

Ecuador

A phytosanitation trial, carried out in the Palma Chávez cocoa plantation within the Pichilingue Experimental Station, provided a site in which an area of cocoa (Area "B") from which brooms had been removed was adjacent to a source area (Area "A") where brooms were still present (see Chapter 13 and Arias, 1990, for more details of the site, and Chapter 7 for Pichilingue climate information). Trees were spaced at 4 x 3 m, and the plantation was divided such that Area "A" comprised 8 rows and Area "B", 17 rows. Seedlings from open-pollinated EET–19, EET–16 and ICS–95 clones were raised for 8 weeks in polythene bags in a greenhouse for protection against witches' broom infection. They were cut back about 10 cm below the apex, and only those plants showing bud activity 2 weeks later were used in the experiment. Seedlings were placed on the ground in Area "B" at 2, 4, 8, 16, 32 and 64 m from the junction with Area "A".

Seedlings were exposed in two seasons, from October 2nd to November 12th, 1987, during the dry season, and from April 12th to May 24th, 1988, during the rains. For both seasons there were 3, initially concurrent, exposure periods of 2, 3 and 6 weeks, with 45 seedlings at each distance, positioned in 3 groups of 15. After exposure, the plants were returned to the greenhouse for evaluation of symptoms, where 50 seedlings had remained as a check on contamination.

Analysis of the Results

The gradient curves obtained from the experiments were analysed using two empirical models, the so–called inverse power law (Gregory, 1968) where $y = ad^{-b}$ (y = disease intensity, d = distance from the source, a = parameter related to the source strength and b = slope of the gradient), and the negative exponential model used by various authors including Frampton *et al.* (1942) and Kiyosawa & Shiyomi (1972) where $y = a^{-bd}$. In the linear form, these models are $\ln y = \ln a - b \ln d$, and $\ln y = \ln a - bd$. Disease was assessed as incidence, and the proportion of seedlings with symptoms was subjected to the multiple infection transformation (Gregory, 1948) prior to regression analysis.

RESULTS AND DISCUSSION

Brazil

The experiment provided data on symptom incidence in both horizontal and vertical planes. A decreasing horizontal gradient was found in the proportion of seedlings with infection for most combinations of source and seedling heights (Figure 12.1). However, when the source was on the ground or at a height of 1 m, there was less infection near the source than farther away on seedlings at 3 m.

Overall, disease incidence was approximately two–thirds less with the source on the ground than when it was elevated, though in making such comparisons, it has to be borne in mind that the exposures at different heights were made at different times, and that basidiocarp production varied from one exposure period to another. As it happened, the basidiocarp totals in the case of the "ground" and "3 m" exposures were similar, and these trials were therefore comparable in that respect. It would seem that brooms on the ground are indeed less strong sources than those in the tree, even when given ideal conditions as in this study by being totally exposed. In natural conditions they might be covered by leaf litter, debris or weeds which could interfere with their wetting and drying, and also with spore dispersal. Moreover, basidiospores falling from basidiocarps on elevated brooms have more time while settling for lateral movement in air currents than spores released from basidiocarps already close to the ground. In addition, air currents are fewer and weaker than those just below the canopy base. Nevertheless, some basidiospores from brooms on the ground obviously were picked up by breezes and convection currents before they could settle on to the litter, and were carried at least to 320 m from the source. Where they occur, productive brooms on the ground should not be ignored as an inoculum source.

When the two models for disease decrease were compared the exponential model, with 20 out of 25 significant regressions, gave a better fit than the power model, and so was chosen for characterising the witches' broom gradients. As mentioned above, source height was confounded with basidiocarp production (and weather), hence no firm conclusions could be drawn regarding their possible individual effects on source "strength" as indicated by values of a. However, for a given source position, there was some tendency for seedlings at higher positions to give smaller values of a as, for example, in the case with the source at 2 m (Table 12.1). When the data for the different vertical positions at each horizontal distance were combined (Table 12.2), the lowest value for a was recorded with the source on the ground, and the highest with the source at 2 m. The number of basidiocarps during the 2 m exposure was 5,618 compared to 2,829 in the 3 m trial.

The slope b of the regression line was variable, but rather flat, with values between

−0.002 and −0.007 in all trials with sources placed above the ground (Table 12.1). For the combined data, it ranged from −0.003 to −0.006, with the 3 m position showing the flattest curve, and the greatest potential for dispersal to bigger distances (Table 12.2).

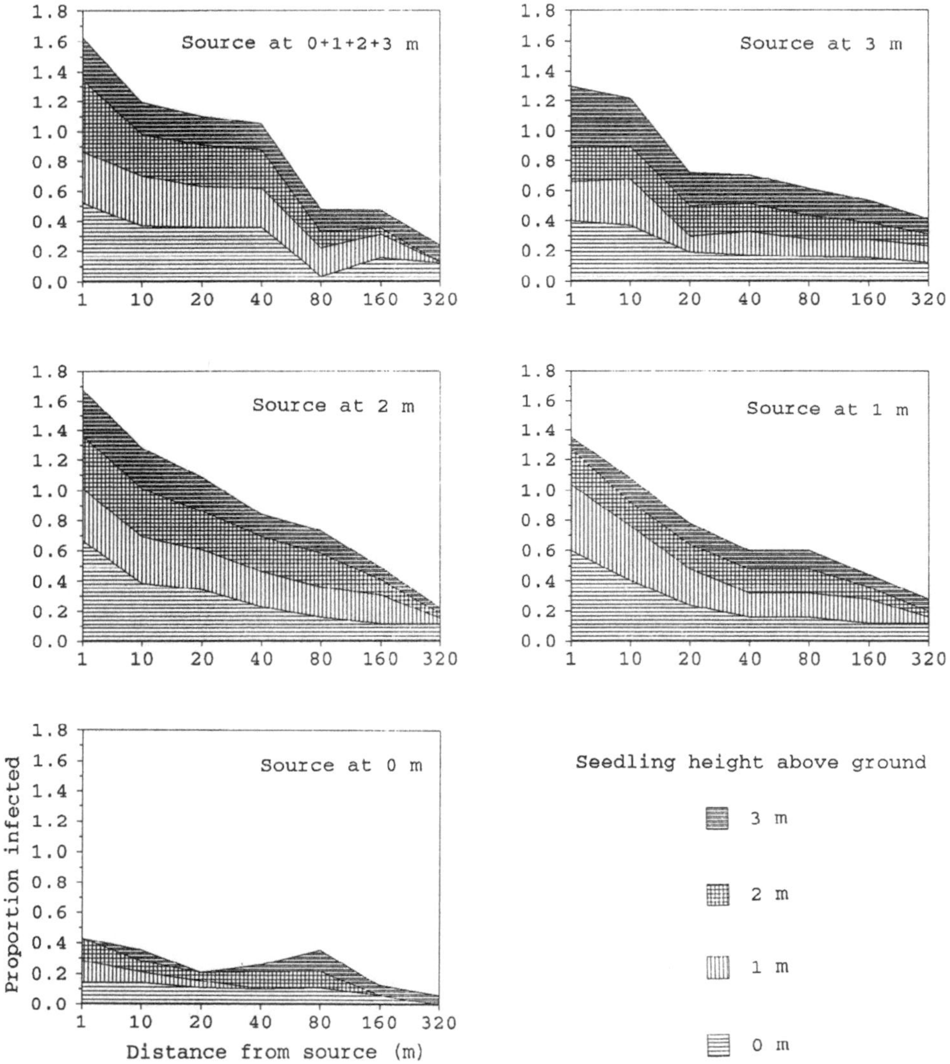

Figure 12.1. Layer graphs showing gradients in the proportion of cocoa seedlings with witches' broom symptoms at different vertical heights, for different source heights. The values on the Y-axis should be divided by four to obtain the overall proportional infection for the summed curves. Rio Negro, Manaus, Brazil.

TABLE 12.1
Parameters to describe horizontal dispersal gradients of witches' broom on seedlings at different heights with the inoculum source also at various heights, after using the multiple infection transformation, and the negative exponential model. Rio Negro, Manaus, Brazil

Position of the source	Seedling exposure height (m)	Parameters estimated			Value of R^2
		ln *a*	*b*		
Ground	0	-0.960	-0.028	**[2]	0.834
	1	-5.260	-0.027		0.422
(3300)[1]	2	-1.844	-0.035	**	0.780
	3	-8.813	-0.026		0.409
1 m	0	-0.974	-0.004	*	0.481
	1	-0.963	-0.007	**	0.847
(6218)	2	-1.525	-0.005	**	0.944
	3	-2.032	-0.002		0.346
2 m	0	-0.760	-0.005	*	0.605
	1	-0.922	-0.007	**	0.925
(5618)	2	-0.935	-0.007	**	0.984
	3	-1.229	-0.007	**	0.938
3 m	0	-1.163	-0.003	*	0.527
	1	-1.570	-0.002		0.305
(2829)	2	-1.413	-0.003	**	0.960
	3	-1.104	-0.004	**	0.735
G+1+2+3	0	-0.975	-0.005		0.281
	1	-0.983	-0.005	**	0.954
(3566)	2	0.414	-0.032	**	0.929
	3	-1.448	-0.003	**	0.718

The intercept *a* is related to the strength of the source, while *b* is the slope of the gradient; [1] number of basidiocarps registered during the exposure; [2] *,** : significant at the level of 0.05 and 0.01, respectively, following the *t*-test.

Presumably, the flat gradients were in part due to the use of a strong source together with very susceptible seedling material. As vegetative brooms in the canopy were shown to be generally the most productive source of basidiocarps in the comparative epidemiology studies (Chapters 5–10), spread from canopy brooms under field conditions is also likely to be considerable, especially in view of the large amounts of inoculum generated by heavily infected trees in the field. Under such conditions, pods which are susceptible for much longer periods (12–14 weeks) than vegetative flushes might show even flatter gradients, or no gradients at all. This aspect is considered in Chapter 13.

TABLE 12.2
Parameters to describe dispersal gradients of *C. perniciosa* for combined seedling exposure heights after using the multiple infection transformation, and the negative exponential model. Rio Negro, Manaus, Brazil

Position of the source	Parameters estimated ln *a*	*b*	Value of R^2
Ground	-2.420	-0.006	0.946
1 m	-1.298	-0.005	0.808
2 m	-0.957	-0.006	0.931
3 m	-1.294	-0.003	0.671
G+1+2+3	-1.047	-0.005	0.784

All the estimates of *b* were significant at the level of 0.01 according to the *t*-test.

TABLE 12.3
Estimates of parameters to describe dispersal gradients of *C. perniciosa* on seedlings exposed for various periods in the Palma Chávez cocoa plantation, Ecuador, using the inverse power law and the negative exponential models, after applying the multiple infection transformation

Exposure period (days)	Parameters estimated ln *a*	*b*	Value of R^2
Inverse power			
14	-0.614	-0.585 *[1]	0.737
21	-0.509	-0.501 **	0.962
42	-0.406	-0.435 **	0.801
Negative exponential			
14	-1.521	-0.024 NS	0.430
21	-1.196	-0.025 **	0.814
42	-1.001	-0.022 *	0.681

[1] *,** : significant at the level of 0.05 and 0.01, respectively, following the *t*-test.

Ecuador

The seedling exposures made in the dry season were completely free from witches' broom infection, despite there being four days during the exposure period with more than 2 mm of rain. In the wet season, nearly 45% of the seedlings were infected close to the source area, while at the maximium distance tested (64 m) the proportion was between 5 and 10% (Figure 12.2). There was a tendency for the longer exposures to give more symptoms.

In contrast to the situation in Brazil, the inverse power model fitted the data better than the negative exponential model, for all lengths of exposure (Table 12.3). The gradient was flatter and the intercept greater as exposure time increased. The curve for the 21 day exposure was also fitted reasonably well by the negative exponential model, and this gave an opportunity for a comparison with the results from Brazil. The slope of -0.025^{**} with $R^2 = 0.814$ was considerably steeper than that for elevated sources in the forest corridor situation in Brazil. Amongst the many differences that must have existed between the two trials, perhaps the presence of the dense cocoa canopy in the plantation in Ecuador would have been foremost as a potential hindrance to the free dispersal of basidiospores at heights of 2–6 m. However, the open space trial of Aragundi *et al.* (1988) in Ecuador, when analysed using the inverse power model, gave estimates for *b* of about −0.50 for 21 day exposures – the same as that found within the Palma Chávez cocoa plantation.

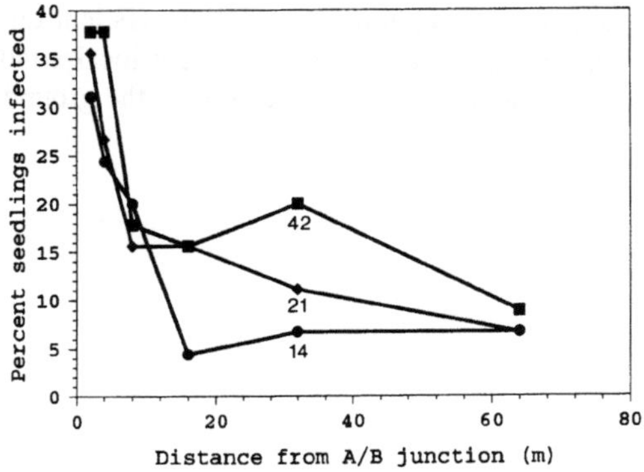

Figure 12.2. Gradients in the percentage infection of cocoa seedlings exposed for various periods on the ground within the Palma Chávez cocoa plantation at different distances from an area with

The fact that decreasing gradients were obtained shows that the careful sanitary pruning practised had been effective within Area "B", and that the amount of backgound infection present during the trial from overlooked brooms and from distant sources did not exceed an amount sufficient to cause 5–10% infection on the seedlings.

IMPLICATIONS FOR DISEASE MANAGEMENT

Considering the results of the trials, and taking them together with observations from the comparative epidemiology studies regarding the prolonged period in which brooms are active during a given season, the quantity of basidiocarps produced, the duration of periods of susceptibility and the number of infection courts available, it is not surprising that the spread of *C. perniciosa* in the field appears to be considerable. Dissemination over tens of kilometres in a single night would be possible theoretically under favourable atmospheric conditions, even though wind speeds in the tropics seldom exceed 3 m per second, and airborne spores are thought to lose their viability in a matter of hours, especially in sunny, dry conditions.

From the point of view of disease management, sanitary pruning which removes the sources of inoculum from all parts of the tree is clearly essential. Basidiospores from basidiocarps on brooms situated high in the tree would seem from the trial in Brazil to have a particularly high potential for distant dispersal, but brooms anywhere on the tree are a threat. Removal to the ground lowers the efficiency of the basidiocarp as a source and, as shown by the comparative epidemiology studies, also reduces, sometimes dramatically, the production of basidiocarps. Nevertheless, basidiocarps on brooms on the ground still act a source of inoculum, albeit a less efficient one, and their removal from the plantation would also have to be considered where local conditions do not result in the rapid covering and decay of diseased material left on the ground.

Chapter 13

COMPARATIVE PHYTOSANITATION STUDIES

A.C. Maddison, T. Andebrhan, F. Aranzazu & R. Silva–Acuña

INTRODUCTION

An understanding of the scope and scale of dispersal of an airborne pathogen such as *Crinipellis perniciosa* is fundamental to its rational management. This is especially true where phytosanitation is being considered as the main element of a management package, as is often the case with witches' broom on cocoa. Before embarking on labour–intensive broom removal, a farmer needs to know whether his efforts will be rendered ineffective by inoculum coming from a neighbour's uncleaned plantation. Similarly, an extension worker or researcher planning a trial of various sanitation practices has to assess if inoculum from uncleaned check plots is likely to interfere with the treated plots, so that adequate guard areas can be included. Otherwise, he runs the risk of underestimating the value of the treatments. To make rational decisions, the farmer, extension worker and researcher would all benefit from information regarding the rate at which disease decreases with increasing distance from a source (the disease gradient), and also on the severity of background infection in a given situation.

The general subject of disease gradients has been reviewed by various authors (Gregory, 1973; Thresh, 1976; Jeger, 1983; Minogue, 1986). Despite its obvious importance to management, there are few reports on gradients and background infection for witches' broom disease in cocoa, and these are restricted to vegetative infections on young trees or seedlings. Baker *et al.* (1941) demonstrated a gradient of vegetative infection in the canopy of young cocoa adjacent to a mature field, and considered the most serious spread to be up to 100 m. Evans & Solorzano (1982) obtained circumstantial evidence from seedling exposure that viable inoculum travels much further, at least up to 800 m, while Aragundi *et al.* (1988) actually recorded gradients of infection on seedlings and spore dispersal gradients in an open area up to 285 m from the nearest source. None of these studies measured gradients of infection on pods, flower cushions or shoots within a mature plantation, or between contiguous farms, and it seems that such information has yet to be published.

Field observations over the years have led to speculation regarding the relative importance of external, or nearby, inoculum in causing pod infections in mature cocoa. On the basis of the results of earlier work on control in Trinidad, including that by Baker & Crowdy (1944), Holliday (1954) considered that witches' broom pod losses were due to infection by basidiospores from local sources, and that they could be controlled fairly

Disease Management in Cocoa: Comparative epidemiology of witches' broom. Edited by S.A. Rudgard, A.C. Maddison and T. Andebrhan. Published in 1993 by Chapman & Hall, London. ISBN 0 412 58190 6

successfully within a field of intensively managed cocoa even if surrounding fields were heavily infected. He also observed that spread of infection from field to field could not be shown clearly by counts of infected pods. Holliday further thought that the failure of sanitation in tall, old cocoa as registered by Thorold (1943) might have been due to brooms left on the ground after pruning or to inefficient cleaning of the inaccessible canopy, rather than to inoculum from other fields. Even in short, intensively managed cocoa in the Brazilian Amazon, experienced pruners still overlooked considerable quantities of infected material on the trees which, according to Evans (1981), could have produced basidiocarps and contributed to the appreciable pod losses recorded after sanitation in that location.

The aims of the studies reported in this chapter were to examine the efficacy of phytosanitation, and to measure gradients of witches' broom disease on pods, flower cushions, and shoots following careful sanitation within mature cocoa plantations in locations with differing disease pressures.

The approach was similar to that in the comparative epidemiology study in as much as a protocol was agreed for use in each of the participating countries and sites: Brazil (2 sites), Colombia (2 sites), Ecuador (1 site) and Venezuela (1 site). The protocol is described below, and then site details and results are presented from each site, together with any departures from the protocol. Other relevant phytosanitation investigations from Ecuador are reported after the protocol studies.

PROTOCOL METHODS

The basis of this Protocol was an initial period of record–taking without treatments in order to calibrate pod production and disease incidence in a block of cocoa. The block was then divided into two contiguous areas. All brooms were removed from one area while the other was left unpruned. Recording continued during the treatment period, and the data on production and disease after sanitation were compared with the calibration data. For the experiment to be sound, the cleaned area had to be isolated on three sides from other cocoa, or any cocoa interfering with the isolation would have to be cleaned to the same high degree of efficiency as the pruned area itself.

After selecting a suitable plantation, a rectangular block of approximately 20 trees by 45 or more trees was marked out and divided into Areas "A" and "B". Area "A" was to be a "source"area, with 10 rows of unpruned trees and Area "B", with at least 35 rows, the cleaned area for detection of gradients. All pods harvested were recorded tree–by–tree in the whole site for about a year, and then brooms were removed from every tree in Area "B", and any trees neighbouring it, except those in Area "A". Cushion and vegetative brooms were counted from six randomly selected trees per row parallel to the source in Area "B", and from one tree per row in Area "A". All brooms were collected and taken away from the plantation.

A second removal count was carried out approximately three months later, with the aim of removing any newly–formed brooms, together with those that had been overlooked inadvertently during the first removal. Counts were again made on the same detailed recording trees in "A" and "B". Soon after, the efficiency of the combined removals was evaluated in "B" by a careful examination of 10 trees randomly selected amongst those with detailed counts. Any brooms still present were counted and classified, and the percentage efficiency for the removal of dry brooms (RED) calculated using the formula:

$$\text{RED} = 100 - ((\text{R/R} + \text{P}_1 + \text{P}_2) \times 100)$$

where R is the number of dry brooms remaining, and P_1 and P_2 are the numbers of brooms removed in the first and second phytosanitary prunings.

In addition to sampling trees in Area "B", 10 trees in those cleaned areas around "B" which might have interfered with its isolation were also checked for broom removal efficiency.

Harvest records were continued, and the broom counts repeated at the end of the treatment year. The data on fruit production and disease, and those for vegetative and cushion broom symptoms were compared with results from the calibration period row by row through Areas "A" and "B". Data from the treatment year were also expressed relative to the first year's data by using the calibration year value for Row 11 (the first row of Area "B") as an arbitrary reference in the following formula:

$$T_n\text{adj.} = (C_{11}/C_n) \times T_n$$

where $T_n$adj. is the adjusted treatment year value for row "n", C_{11} is the calibration year value for row 11, C_n is the calibration year value for row "n", and T_n is the treatment year value for row "n".

Percentage infection data were transformed using the multiple infection transformation (Gregory, 1948). The disease data were then plotted on linear and logarithmic scales to help determine the form of gradients and their respective background points, if present (Gregory, 1973). The negative exponential model (Frampton *et al.*, 1942; Kiyosawa & Shiyomi, 1972) where $y = a^{-bd}$ (linear form $\ln y = \ln a - bd$; y = disease intensity, d = distance from the source, a = intercept on disease axis and b = slope of the gradient), was generally more appropriate than the inverse power model. Regressions were carried out of ln (natural logarithm) of disease against distance in metres from the nominal "A"/"B" canopy junction, including row 10 of Area "A" (or the mean of the values for the single trees in rows 6–10 in the case of broom count data).

Grouping of the row data was considered for smoothing row to row variation, but in certain cases it obscured changes in the curves and tended to flatten gradients spuriously.

In some sites, basidiocarp production was monitored weekly on vegetative brooms suspended from five trees in Area "A", with two types of brooms in each tree – brooms selected at random, and those already known to be productive. There was a total of 100 brooms in each tree.

SITE DESCRIPTIONS AND RESULTS

Brazil

Site Descriptions

Altamira. The site near Altamira (Pará) was the same as that used for the comparative epidemiology experiment (for description see Chapter 5), except that the whole farm was employed, with 118 trees in the 10 rows of Area "A", and 591 trees in Area "B". It was an isolated, unshaded commercial farm planted with hybrid seed at 3 x 3 m, and surrounded by pasture and secondary forest (Figure 13.1). The nearest cocoa planting was 300 m away, 19 ha in extent, and characterised by heavy overhead shade, appreciable canopy infection, but low pod infection by witches' broom.

Ouro Preto. In Ouro Preto (Rondonia), an abandoned 12 years old farm was used, with trees from crosses of CEPLAC clones planted at 2.5 x 2.5 m. Previously, inadequate formation pruning had allowed excessive vertical growth, to an average height of 6 m. Phytosanitary pruning had been done rarely, and pod harvest was the only current activity. The one hectare farm was divided into an Area "A" with 12 rows and 146 trees, Area "B" with 38 rows and 414 trees and a remaining guard area beside Area "B" with 560 trees (Figure 13.1). The site was isolated, and surrounded by pasture; the nearest cocoa farm was about 500 m distant.

General. The experiments ran from November 1987 to October 1989 and were conducted according to the Protocol, except that in Altamira the broom count data at the end of the treatment year were from all trees, and not just from those recorded at the end of calibration period. The phytosanitation was done in August/September, and the repass in November/December, 1988. Rainfall data were taken from the CEPLAC experimental station 5–6 km from the sites.

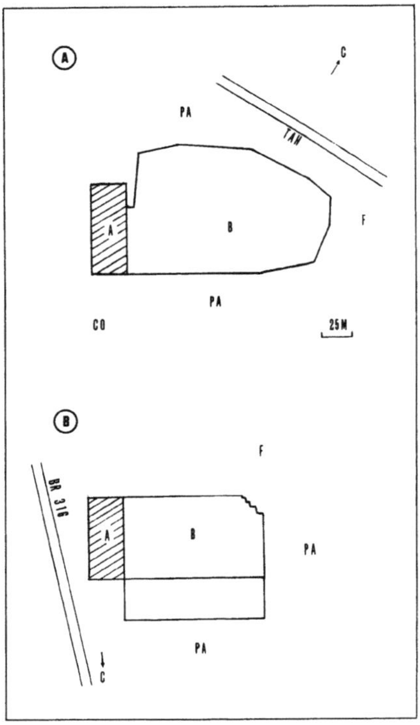

Figure 13.1. Plans of the Brazilian Comparative Phytosanitation experimental sites in Altamira (A) and Ouro Preto (B); C, cocoa; CO, coffee; F, secondary forest; PA, pasture; TAH, Trans-Amazonia Highway. Hatching indicates cocoa lacking phytosanitation in the treatment year.

Results

Rainfall. The rainfall in the calibration year was similar to the long-term average in both sites with totals of 1,900 and 1,941 mm, but the treatment year was very wet

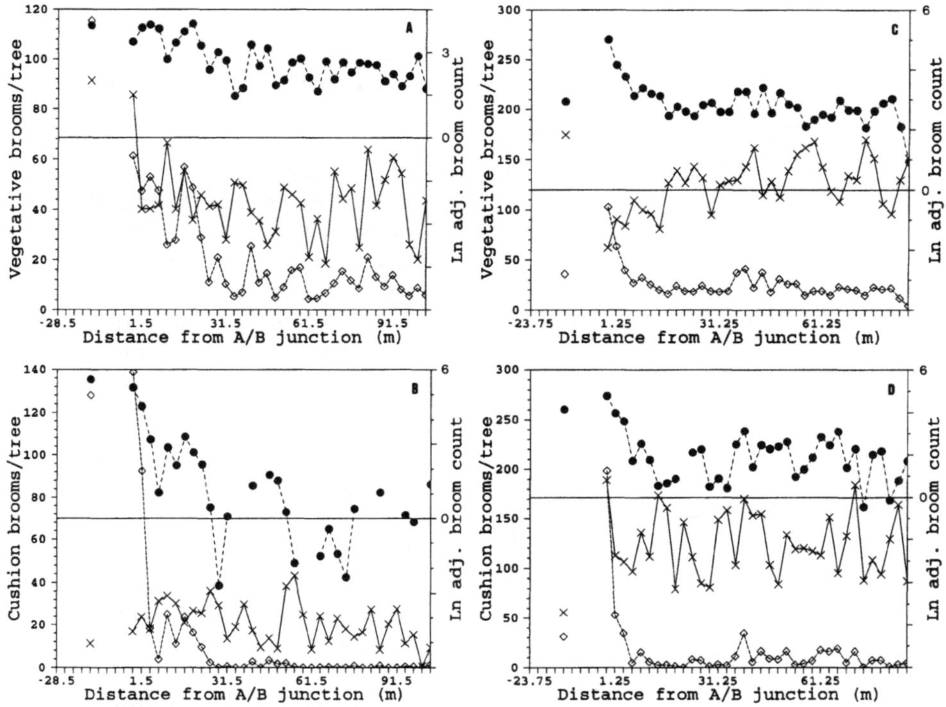

Figure 13.2. Witches' broom incidence row–by–row on shoots and flower cushions in Altamira (A,B) and Ouro Preto (C,D), Brazil, in the calibration and treatment years. ×, calibration; ◊, treatment year; ●, ln adjusted treatment year. The values in Area "A" are each means of five trees (one per row), while those in "B" are for the six trees sampled per row.

with 2,215 mm in Altamira and 2,243 mm in Ouro Preto.

Infectionon shoots and flower cushions. Prior to sanitation, infections on shoots and cushions were distributed more or less uniformly throughout both sites, though variability from row to row in the six–tree means was appreciable (Figure 13.2). After sanitation, the incidence of witches' broom fell dramatically in Area "B". There were decreasing gradients of symptoms away from Area "A" on both flower cushions and shoots. At Altamira, the decrease in cushion symptoms continued with increasing distance, so that virtually no cushion brooms were recorded beyond 40 m from the "A"/"B" junction, while at Ouro Preto some cushion symptoms were found up to the greatest distance tested. The numbers of symptoms on shoots did not fall to zero at either Altamira or Ouro Preto, but levelled out at approximately 40% and 20% of the values for the calibration year in the two sites, respectively (Figure 13.2).

When the natural logarithms of the adjusted shoot and cushion broom data were plotted against distance (Figure 13.2), the curves showed two components, an initial

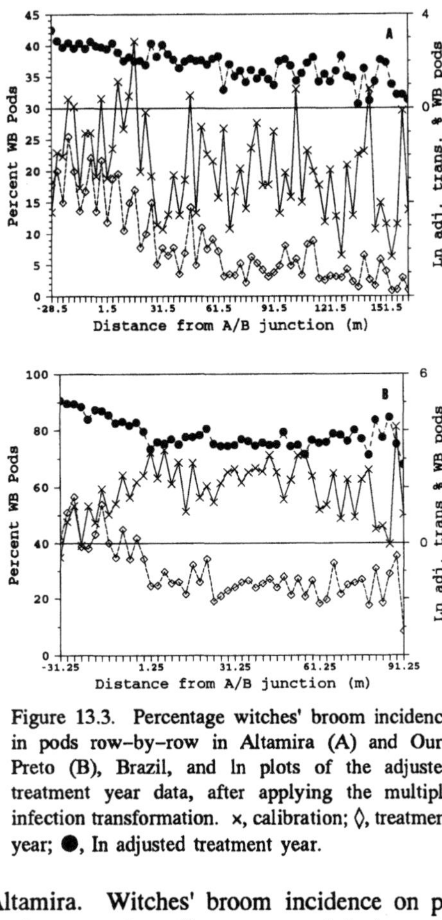

Figure 13.3. Percentage witches' broom incidence in pods row-by-row in Altamira (A) and Ouro Preto (B), Brazil, and ln plots of the adjusted treatment year data, after applying the multiple infection transformation. ×, calibration; ◊, treatment year; ●, ln adjusted treatment year.

exponential (more or less straight line) decrease, followed by an horizontal "plateau", which presumably represented background contamination from missed brooms within the plantation, together with inoculum coming from afar. The "background point" was reached farther away at Altamira (40 m from the "A"/"B" junction) than at Ouro Preto (20–25 m). Regressions of the ln adjusted values against distance were significant, the slopes being generally less steep at Altamira than at Ouro Preto, and less steep for shoots than for cushions at both sites (Table 13.1, Discussion section).

The evaluation of pruning efficiency carried out after the repass at the end of the calibration year showed that 95–96% of the brooms had been removed from the sample trees in both sites. Most of the missed diseased objects were green brooms.

Pod production and disease. The high variability from tree to tree, which is typical of cocoa grown from mixed seed (Griep & Lima, 1988), gave rise to considerable variation in pod production from row to row in the calibration year. Also, there was a tendency for production to be appreciably more at one end of the plot than the other, particularly at Altamira. Witches' broom incidence on pods fluctuated from row to row, but no clear environmental gradients emerged during calibration at either site (Figure 13.3). Pod losses to witches' broom were much lower (20%) at Altamira during calibration than at Ouro Preto (60%).

After phytosanitation, there was a gradient fitted by the negative exponential model at Altamira, with the percentage of pods affected by witches' broom falling to a mean of about 5% at the far end of "B". Sanitation reduced pod infection in Area "B" at Ouro Preto to a mean of 25%, but there was no gradient, except perhaps between the two rows adjacent to "A" (Figure 13.3). In addition to lowering pod loss, sanitation was also associated with an increase in pod production at Ouro Preto, where pod numbers increased by 16.3% in Area "A", but by 42.1% in Area "B". A similarly large effect was not observed at Altamira, where pod production fell in the treatment year by 13.6% in Area "A" and by 11.1% in Area "B" (Figure 13.16).

Colombia

Site descriptions

Experiments were carried out in two important cocoa areas with different climates, one near Manizales, Caldas, at 1,020 m a.s.l. in the marginal lower coffee zone and the other in the municipality of Guamal, Meta, at 200 m a.s.l. in wet tropical forest on the Llanos Orientales (Eastern Plains). The sites were established by the Cocoa Programme of the Colombian Agriculture Institute (Instituto Colombiano Agropecuario) in January 1988 and ran for two years.

Manizales. The experiment was sited in the "Montelindo" Farm, property of the University of Caldas, in an area adjoining the old IWBP comparative epidemiology experiment (see Chapter 6). All the cocoa in the experimental area was of clone SC–6, 27 years old, planted at 4 x 4 m triangular, under very irregular shade. Area "A" had 10 rows of trees and Area "B" 35 rows, with a nominal 20 trees per row. There was cocoa adjacent on two sides, woodland on the third and soya on the fourth (Figure 13.4). Phytosanitation had been practised from time to time in the cocoa.

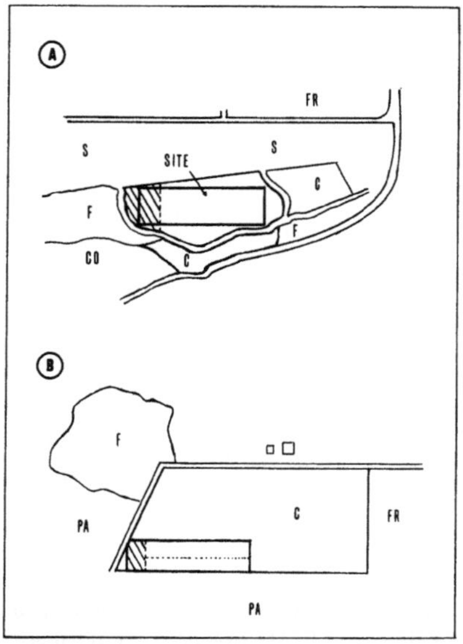

Figure 13.4. Plans of the Montelindo, Manizales (A) and Guamal, Llanos (B) experimental sites in Colombia showing the immediate surroundings. C, cocoa; CO, coffee; F, forest; FR, fruit trees; PA, pastures; S, soya. Hatching indicates cocoa lacking phytosanitation in the treatment year.

Guamal. This site was established in the farm, "Los Alamos", in hybrid seedling cocoa planted at 4 x 4 m triangular, 18 years old, with irregular shade. The site occupied the corner of a larger plantation (Figure 13.4), with 10 rows in Area "A" and 35 in "B". Area "B" was divided longitudinally to accommodate a superimposed fungicide trial, which is not reported here.

Methodology

In both Manizales and Guamal, there was a deviation from the Protocol in that brooms were removed and counted from sample trees in February 1988, at the beginning of the experiment. At other sites the first broom removal was at the end of the calibration period. Thereafter in Colombia the sample trees, one per row in Area "A" and six per row in Area "B", had brooms removed and counted at six–monthly intervals from February 1988 until February 1990. The remaining, non–sample trees in Area "B" (amounting to 58% of the total in Manizales and 28% in Guamal) did not receive phytosanitation in February 1988, but brooms

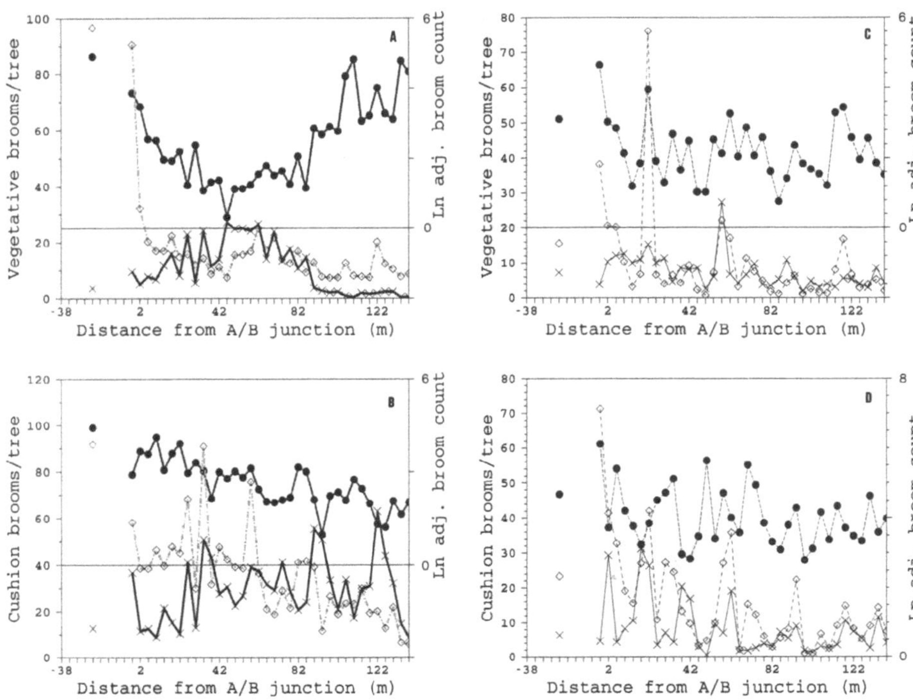

Figure 13.5. Witches' broom incidence row by row on shoots and flower cushions in Manizales (A,B), and Guamal (C,D), Colombia. ×, reference count; ◊, 1989/90 count; ●, 1989/90 ln adjusted treatment year count.

were removed at the same time as for the sample trees from August 1988, onwards. This procedure meant that there was no calibration year as such in which brooms were allowed to accumulate naturally. The broom data of February 1988 do provide an initial calibration of sorts, though it is not known exactly how long the plantations had been without sanitation prior to that of February.

The results of the broom removal in August 1988 might have reflected the infection which occurred on shoots and flower cushions of cleaned trees when neighbouring trees in Area "B" had not been cleaned; and those from February 1989, the situation when all trees were cleaned. However, as green brooms continued to appear immediately after the August 1988 removal from infections that had occurred in the February–August period, the February 1989 data are in fact a combination of infections from two sanitation regimes. For this reason the August results are combined with those of the following February in the analysis. Evaluations of the thoroughness of the sanitation were made in March or April 1989, in Area "B" and in the neighbouring cocoa.

Because of the presence of moniliasis *(Moniliophthora roreri)* in the Manizales site, diseased pods were removed weekly, and healthy ripe pods every two weeks. The pod harvest data were summarised in two periods; from March to November 1988, when pods in Area "B" had been under higher inoculum pressure, and from December 1988 to June 1990, when all of "B" had been cleaned. At Guamal harvesting was fortnightly, but uncertainties over the validity of the data during the peak of pod loss in October and November 1989 led to the results for pods being omitted.

Results

Rainfall and basidiocarps. At both sites, 1989 was wetter than 1988. Rainfall amounted to 2,180 mm compared with 1,964 mm at Manizales, while at Guamal the respective figures were 2,687 mm (up to week 49) and 2,012 mm. Basidiocarp production followed the patterns seen in the comparative epidemiology study (Chapter 6), with a decrease at Manizales during the first quarter of the year, and few or no basidiocarps at Guamal in the January/February dry season. Green brooms on the point of drying produced basidiocarps after a minimum of 24 weeks at Manizales, compared with 17 weeks at Guamal.

Infection on shoots and flower cushions. Even in the clonal material, there was considerable variation in the number of witches' brooms from tree to tree and row to row at the beginning of the experiment. Tree size and vigour varied, as did overhead shading, and there was a distinct environmental gradient in shoot symptoms at Manizales (Figure 13.5). At both sites, the lack of sanitation in Area "A" led to a considerable increase in the number of shoot and cushion brooms recorded in August 1988 and February 1989. In Area "B", the adjusted data showed a decreasing gradient in symptoms away from Area "A" at Manizales, but not at Guamal. The decrease was less abrupt for cushion symptoms than for those on shoots, which additionally showed a marked increase in abundance beyond Row 30, in the sector where infection had been minimal in February 1988.

The evaluations of pruning efficiency in March 1989 showed that on average at Manizales there were 6.6 dry brooms and 4.6 green brooms remaining per tree in Area "B" after the February sanitation. At Guamal, despite the smaller numbers of brooms initially, the corresponding figures were 10.8 dry brooms and 2.3 green brooms. In the cocoa alongside Area "B", at both Manizales and Guamal sampled trees each had 12–14 dry brooms after sanitation.

Pod production and disease. Harvests at Manizales during the reference period March–November, 1988, showed no marked gradient in productivity within the site parallel to Area "A", but the percentage of pods with witches' broom was lower at the distal end of Area "B" (Figure 13.6). After the repeated sanitation in Area "B", pod loss was reduced from 20.0% to 11.7%, while in Area "A" it remained at about 27%. The reduction occurred almost throughout "B", with the exception of the last few rows where "B" was bordered by other cocoa. There was a 'step–down' between "A" and "B", but no marked exponential decrease away from "A" in witches' broom symptoms on pods. Pod production per tree per month increased by 69% in Area "B" from December 1988, compared to only 18% in Area "A" (Figure 13.16).

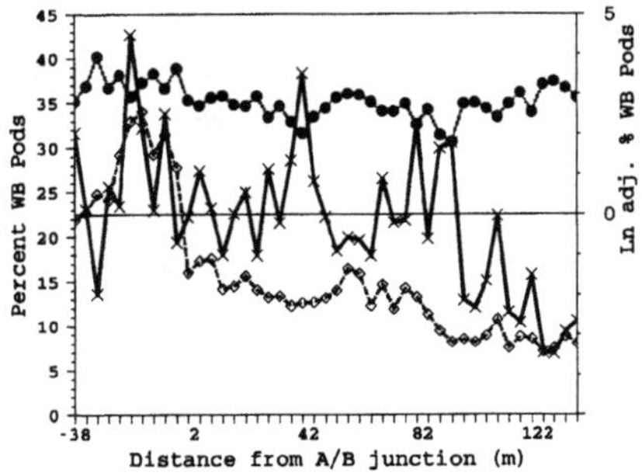

Figure 13.6. Percentage witches' broom incidence in pods row by row, and ln percentage witches' broom adjusted relative to 1988, after applying the multiple infection transformation; Montelindo, Manizales, Colombia. ×, March–November, 1988; ◊, December 1988 - June, 1990; ●, ln adjusted values.

Ecuador

Site description

The site chosen was in Lote Juban, Hacienda Balao Chico, 20 km south of Naranjal, Guayas Province, Ecuador (79° 35'W and 02°45'S) at an altitude of 6 m a.s.l., on the narrow coastal plain adjacent to the Andes south–east of Guayaquil (Lara, 1991). The climate is similar to that of Pichilingue, but somewhat drier and even more variable, with an average rainfall (Hacienda Pechichal, Naranjal, 1964–81) of 1,053 ± 686 mm. Like Pichilingue, the dry season (June–December) has periods with heavy mists and occasional drizzle, known locally as "garua". Channel/flood irrigation is commonly used in this zone for commercial banana and cocoa plantations.

Lote Juban was planted in 1980 at 3 x 3 m, using open pollinated seed from Nacional/Trinitario hybrids. Heavy *Erythrina* sp. shade, at 6 x 6 m in certain areas, was thinned in 1987 to 12 x 6 m. The cocoa, which had not received formation or phytosanitary pruning previously was tall, up to 7 m, with several storeys. Further upward growth was prevented when the area was taken over in March 1988. The site, on flat alluvial soils, had pasture or bananas on three sides and on the fourth more cocoa which extended to about 30 ha (Figure 13.7). It was irrigated in July and August, 1988.

Pod harvests were made using the scheme described for the Pichilingue comparative epidemiology studies (Chapter 7), with the calibration period running from March to December 1988, and the data year from January to December 1989. The first broom removal took place in September 1988, with a follow–up removal in November /December, and an evaluation of the efficiency of the removals in February 1989. The 1989 broom removals were done in the same months as in 1988, with the evaluation in January 1990.

As an additional treatment, fungicide (Cobre Nordox 50, 58% cuprous oxide) was applied every two weeks from week 50, 1989, until week 11, 1990, to the eastern half of Area "B" using a "Nuvola" motorised knapsack sprayer. Approximately 2 g of active ingredient were applied per tree, per application, initially in 100 ml and later in 150 ml of water per tree to improve the coverage.

Basidiocarp counts and meteorological data were recorded, respectively from weeks 40 and 48 of 1988, with an additional set of brooms placed on the ground. Basidiocarp counts often had to be made on a fixed day of the week, not necessarily after rain.

Results

Climate and basidiocarp production. In the calibration year, 1988, the total rainfall of 979 mm (Hda. Pechichal, Naranjal) was close to the long-term mean, while 1989 was a wetter than average year with 1,683 mm recorded in the site. More than 1,000 mm fell in February and March 1989 alone, and this appears to have suppressed pod production. Night-time winds, if they occurred, generally were light and from the north to north-east; calm often prevailed around dawn, and early morning breezes usually came from the south-west.

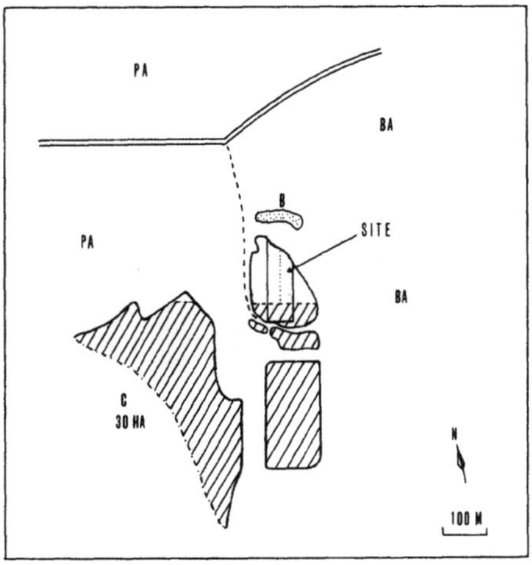

Figure 13.7. Plan showing the Balao Chico site, Ecuador, and surroundings. B, bamboo; BA, bananas; C, cocoa; PA, pasture. Hatching indicates cocoa lacking phytosanitation in the treatment year.

Basidiocarp formation did not correlate well with the rainfall total (R^2 = 0.038 NS), because small quantities of rain and periods of drizzle were sufficient to stimulate production, especially after a dry period. Thus basidiocarps were present in the main wet season, but there were also strong peaks in June/July and September/October 1989. Production was appreciable in November/December 1988, but not in 1989.

Infection on shoots and flower cushions. There was a strong environmental gradient for both shoot and flower cushion symptoms at the end of the calibration period. Symptoms were fewer in the more northerly half of Area "B", farthest from "A" (Figure 13.8). After sanitation, there was little change in the number or distribution of the sparse symptoms on shoots within "B". For cushion symptoms, there was a suggestion of a decreasing gradient up to about 75 m from Area "A", but also an increase at the northerly end.

The evaluation of broom removal efficiency showed that on average only one broom per tree remained in Area "B" and adjacent cocoa after the second removal, in both 1988 and 1989. Removal from the ground was more thorough in the Year 2 with 0.4 brooms per tree remaining, compared to 2.2 in the Year 1.

Pod production and disease. At the end of the calibration period, pod production and disease incidence per row were variable, but without showing any consistent trend parallel to the "A"/"B" junction, except for lower incidence of both witches' broom and moniliasis on pods in the ten most northerly rows of Area "B". After sanitation, there was no indication of any additional decreasing gradient in the percentage of pods with witches' broom in Area "B" (Figure 13.9). When a peak of pod infection was considered, for example weeks 18–28 of

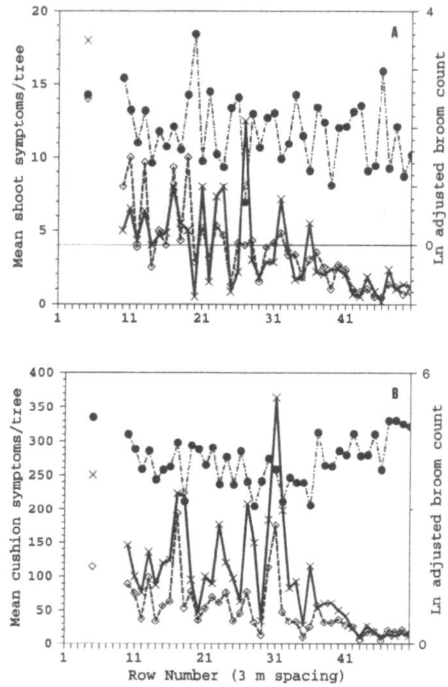

Figure 13.8. Incidence of witches' broom on (A) shoots and (B) flower cushions in the Balao Chico site, Ecuador, during the calibration (×) and treatment year (◊); ●, ln adjusted broom counts.

1989, the overall proportion of pods with witches' broom was higher than for the whole year, but the slope of the curve differed relatively little from that in the calibration year (Figure 13.9).

Although witches' broom was not the only cause of pod loss, the damage caused by moniliasis, *Phytophthora* pod rot and a mirid (*Monalonion* sp.) did not distort greatly the row–by–row percentage figures. The use of "witches' broom + healthy" as the denominator rather than "total fruits" when calculating the percentage witches' broom, did not change materially the curves obtained; the correlation between the two percentages was high in all cases (R^2 >0.90).

Pod production was generally lower in 1989 than during calibration, when the mean production for both "A" and "B" was about 24 pods per tree. Area "B" showed a smaller reduction in 1989 to 17.1 fruits/tree, compared to 13.9 in "A" (Figure 13.16). Witches' broom incidence on pods was up by 75% on the 1988 value in "A", while the increase in "B" was only 52%.

The substantial border area to the west of Area "B" showed a more uniform distribution of witches' broom fruits during the calibration period than had the central block, and in 1989 a slight gradient falling away from the source was noticeable with $b = -0.0050^{**}$, compared to -0.0011^{NS} in 1988. The displacement from the source necessary for a 25% reduction in the proportion of fruits diseased was approximately 70 m.

Effect of fungicide applications. The year of the fungicide applications was a year with little witches' broom in the Balao Chico area in general. For the period up to week 38 1990, Area "A" showed a loss of only 8.0% of total pod production to witches' broom. In Area "B", the unsprayed part lost 5.1%, and that with fungicide 3.6%. As can be seen from Figure 13.9, the effect of the fungicide was noticeable only at distances of more than 60 m from Area "A".

Venezuela

Site description. The site had the same location as the comparative epidemiology experiment (see Chapter 10 for details) in an isolated 5 ha plantation of seedling criollo cocoa in the

"Primavera" farm, El Piñal, Isla de Betancourt, Táchira State, at 7°32'00" N 71°57'18" W and 270 m a.s.l. The comparative epidemiology site was contiguous with Area "A" (162 trees), while Area "B" (582 trees) was bordered by plantains, cassava and pastures (Figure 13.10). Cocoa outside the IWBP experiments was weeded and pruned by the farmer twice in the year, once in April before the rains (annual mean of 2,834 mm) and again at the beginning of the dry season in December. Brooms were removed from the trees and left on the ground.

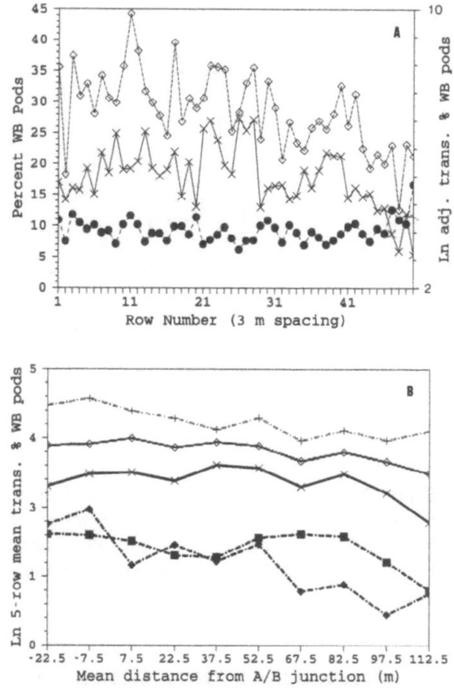

Figure 13.9. (A) Percentage witches' broom incidence in pods row by row in Balao Chico, Ecuador, during calibration year (×) and treatment year (◊), and ln adjusted transformed values (●); (B) semi-ln plots of the 5-row means of adjusted transformed percentage witches' broom in pods for wk 18–28, 1989 (+), and adjusted values for sprayed (♦) and unsprayed (■) half of "B" in 1990.

The cocoa was spaced at approximately 4 x 4 m under irregular *Erythrina* sp. shade at about 7 x 7 m. Rows in the original planting were not parallel so, for the purposes of summarising harvest data, Area "B" was divided into 15 sections each of 10m, which were recorded every two weeks from July 1988. Wilted cherelles were included in the "others" category during harvests, hence pod production and percentage pods with witches' broom were calculated using "ripe pods + witches' broom pods".

Brooms were removed from the ten sample trees in Area "A" and from all trees in Area "B" at the end of August, beginning of September 1989 after the year's calibration. In "B", counts were made of the brooms on 14 trees in each of the 10 m sections. Evaluations of the green brooms apppearing on two trees in Area "A" and 20 trees in Area "B" were made two weeks after the removals. In 1990, the removals with counts were made in the second half of May, and the evaluation of efficiency at the beginning of June. Basidiocarps were counted every two weeks on the sample sets of brooms, on fixed days not necessarily after rain.

Results

Climate and basidiocarp production. In 1989, April was drier than usual and the annual total rainfall was 2,272 mm, compared to 2,621 mm in 1990. Basidiocarp production was sporadic and low, as in Set 2 of the comparative epidemiology study (Chapter 10).

Infection on shoots and flower cushions. The mean number of brooms on shoots and

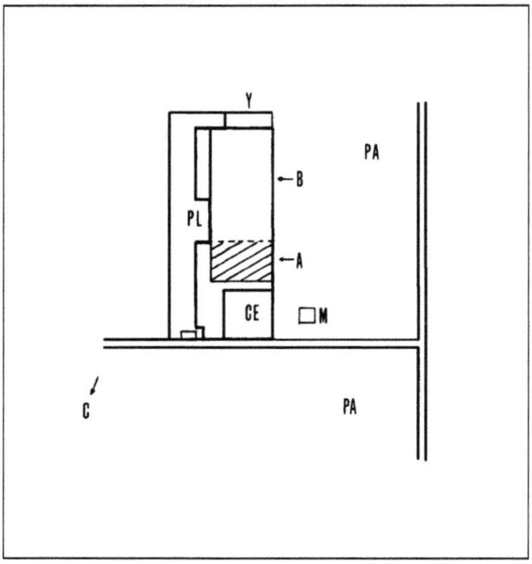

Figure 13.10. Plan of the comparative phytosanitation site, El Piñal, Venezuela; C, cocoa; CE, comparative epidemiology site; M, meteorology site; PA, pasture; PL, plantains; Y, yucca. Hatching indicates cocoa lacking phytosanitation in the treatment year.

cushions varied considerably from tree to tree and section to section at the end of the calibration period in 1989 (Figure 13.11). Broom incidence generally fell in the year after sanitation, but Area "B" showed a bigger reduction than Area "A", even close to it, especially for witches' broom on shoots. With such vegetative brooms, there was a distinct "step down" in passing from "A" to the first section of "B", and an irregular decrease up to 105 m from "A". Cushion broom counts in "B" were less consistent from section to section, showing a decrease close to "A", but irregular high values beyond 70 m from "A". Taken overall, between 1989 and 1990 the mean values in "A" fell from 87.1 to 52.3 vegetative brooms/tree and from 20.5 to 10.5 for cushion brooms. In "B" the fall was from 62.8 to 7.2 vegetative brooms/tree and from 18.1 to 6.4 cushion brooms.

Regression analysis of the ln disease data versus distance from Area "A" gave no significant lines for cushion brooms, where there were several zero values in the Year 2. With vegetative brooms up to 105 m from "A", the calculated slope for the adjusted data was -0.0174 with $R^2 = 0.44$.

With regard to broom removal efficiency, no dry brooms were recorded on the sample trees after sanitation, but newly emerged green brooms were found, particularly after the first removal in August/September which occurred when brooms were still appearing.

Pod production and disease. Pod production in Area "A" fell slightly in 1990, while in "B" it remained at a similar level to that in 1989 (Figure 13.16). The proportion of witches' broom pods, which had been about 75% of ripe pods plus witches' broom pods in both areas in 1989, increased to 80.6% in "A" but fell to 69.3% in "B". In terms of the absolute numbers of ripe pods harvested, this meant a fall of 25% in "A" compared to an increase of 16.7% in "B", relative to the 1989 values (Figure 13.16). Data from the comparative epidemiology experiment in the same plantation showed that there were very few pods affected by *Phytophthora* pod rot or other diseases in 1989; ripe pods and witches' broom pods accounted for virtually the entire harvest.

When looked at section by section, there was an environmental gradient in production in Area "B" in both years, decreasing from about 70 pods per tree near to "A", to 35 per tree at the distal end. The losses to witches' broom were high throughout the site in 1989, and the small reduction in "B" in 1990 did not show a gradient away from "A" (Figure 13.12). The adjusted 1990 data did show a very slight gradient, with $b = -0.0009^*$.

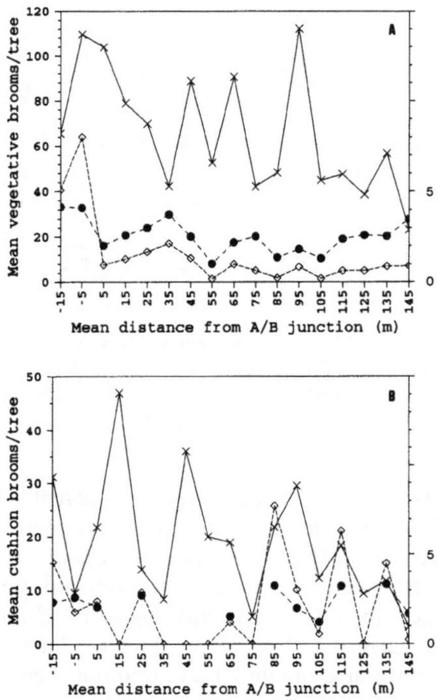

Figure 13.11. Mean numbers of (A) shoot and (B) cushion brooms per tree per section in El Piñal, Venezuela, during the calibration (×) and treatment year (◊), and the ln adjusted broom counts (●). Missing ln counts indicate zeros.

Ecuador Phytosanitation: Related Studies

Introduction

The experiments briefly reported here were started in Ecuador towards the end of 1986, to investigate the effects of sanitation in small plots of relatively isolated cocoa, both within the Pichilingue Station, and in farmers' traditionally tall cocoa (Arias, 1990). The Pichilingue site was a forerunner of the protocol phytosanitation study described above, with an uncleaned source area adjacent to a cleaned area, while in the farmers' cocoa brooms were removed from all the trees in the planting.

Sites and methods

Pichilingue Experiment Station. The plot used, "Palma Chávez", was 250 m from the nearest cocoa, and originally comprised 25 by 25 trees. It was planted in 1969 at 4 x 3 m, under *Inga edulis* shade, but this had since disappeared. The trees were from seedlings of two interclonal hybrids, IMC–67 x Sca–6, and EET–19 x Sil–1. The average height of the surviving 546 trees was 5.15 m in 1989, and they had received formation, maintenance and some phytosanitary pruning throughout their lives. Area "A" occupied 8 rows and Area "B" 17 rows.

Broom removal from Area "B" was carried out towards the end of the dry season in 1986; brooms were collected row by row, parallel to the source, allowed to dry and then weighed by row before being destroyed. An evaluation of the 1986 broom removal was made on March 24 1987, rather a long time after the sanitation, and without knowing how many brooms were present originally. In 1987 and subsequent years, brooms were removed very conscientiously in September/October and again in November/ December, with several passes over the trees during each removal. Counts of brooms were made tree by tree, and air dry weights were measured row by row. The efficiency of the removal process was assessed in

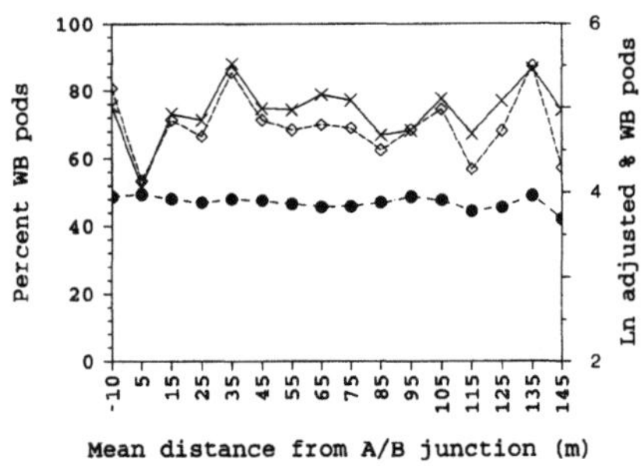

Figure 13.12. Percentage witches' broom pods per section in El Piñal, Venezuela, during the calibration (×) and treatment year (◊), and ln adjusted percentage witches' broom pods (●).

the years 1988–90 by counting the dry brooms that remained on a sample of 36 trees in late January or early February – after the second removal.

Harvest data were collected from week 4, 1987 as described above for the Balao Chico site, with the exception that as from week 41, 1987, diseased fruits were removed weekly rather than every 2 weeks as in the case of ripe pods, in an attempt to reduce the losses of fruits to moniliasis.

Traditional farmers' cocoa. A small plot of 16 yr–old traditional cocoa was chosen, separated by about 300 m from other cocoa, at the Finca San Juan, 35 km north–east of Pichilingue. The cocoa was on average 8 m tall, and of mixed parentage, planted at about 3 m square on level ground, with some permanent shade trees interspersed. Brooms were removed, after counting on 10 sample trees, from all of the 400 trees in November/December 1986, and again in 1987. Pod harvests were carried out in the standard way every 2 weeks on the 10 sample trees, and on 40 neighbouring trees, with unidentified diseased fruit split and left in the plantation for subsequent detection of moniliasis. Two plots of 50 trees each were also defined in an extensive area of similar tall, traditional cocoa, which was about 300 m from the first group of trees in the Finca Santa Aurelia. This cocoa was about 30 years old and, like San Juan, previously unpruned. Brooms were removed from trees in one plot, while the second was left uncleaned, like the rest of the plantation. Evaluation of pruning efficiency was delayed after the first removal, not taking place until May 1987, when old brooms alone were counted. The 1988 evaluation took place in January, 1989, again on the 10 detailed recording trees.

Results

Palma Chavez – infection on shoot and flower cushions. The broom weights per row obtained at the end of 1986 provided a calibration of combined shoot and cushion broom incidence in Area "B", and revealed a strong "environmental" gradient prior to sanitation (Figure 13.13). After sanitation, witches' broom incidence was lower in "B", especially in 1989 and 1990 when "A" was also cleaned. If expressed as a percentage of the 1986 values, there was a decreasing gradient away from Area "A" in 1987 and 1988, which levelled off 20–40 m from the "A"/"B" junction (Figure 13.13). This gradient disappeared when "A" was also cleaned. Despite careful sanitation of the entire plantation, with only 1.1 and 0.4 dry brooms per tree remaining on average after the 1988 and 1989 removals, means of 40.4 and 41.3 new brooms per tree were recorded in 1989 and 1990, respectively.

The slopes of the logarithmic part of the curve (up to 34 m), obtained from the regression of ln broom weight as a percentage of 1986, versus distance from "A" in metres were -0.032^{*} (R^2 = 0.606) and -0.051^{**} (R^2 = 0.947) for 1987 and 1988, respectively, compared with -0.023^{**} ($R^2 = 0.694$) and -0.027^{**} ($R^2 = 0.799$) in 1989 and 1990.

Palma Chavez – witches' broom infection of pods. No pre–treatment calibration for pod production or pod disease incidence was possible within the thesis time–frame, but the results for 1989 and 1990, when brooms had been removed from both "A" and "B", act as a post–treatment calibration. They showed that pod production tended to be greater at the distal end of Area "B", while the loss to witches' broom tended to be less. There was a slight downward, environmental, gradient in disease incidence in Area "B" with increasing distance from "A" in both 1989 and 1990 (Figure 13.14).

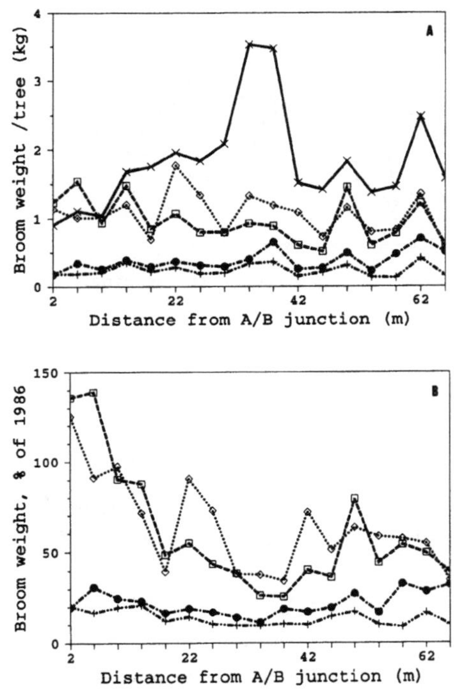

Figure 13.13. (A) Row by row dry broom weights in Area "B" of the Palma Chávez plantation, Pichilingue, Ecuador, 1986–90. (B) ln broom weights as a percentage of the 1986 reference. [×, 1986; ◊, 1987; □, 1988; ●, 1989; +, 1990]

The removal of brooms from Area "B" did not appear to bring about a marked decrease in the percentage of fruits affected by witches' broom as one passed from Area "A" to Area "B", either in 1987 following the less efficient broom removal at the end of 1986 (29.7 dry brooms/tree remaining in March, 1987), or in 1988 after the 92.3% removal of 1987 (15.7 dry brooms/tree remaining). However, in 1989 and 1990, after Area "A" was cleaned too, and sanitation was more than 99% efficient, there was a reduction in fruit loss in Area "A", relative to that in Area "B", and to previous years (Figure 13.14). When the data for 1987, 1988 and 1989 are expressed as a ratio of those for 1990, the three curves are similar in Area "B", while in Area "A" those for 1987 and 1988 are consistently somewhat above 1989 (Figure 13.14). This suggests that the sanitation did reduce pod loss generally in Area "B" in 1987 and 1988, though without showing a marked gradient. In 1990, pod losses to witches' broom throughout the plantation were less than 10%, though it has to be borne in mind that climate, as well as sanitation, was governing disease incidence from year to year.

The overall production of fruits increased in Area "A" by 27.1% and 24.5% in 1989 and 1990, respectively, relative to 1987, while the corresponding increases in Area "B" were 10.8 and 3.9%. Presumably similar beneficial effects of broom removal on pod production had already taken effect in Area "B" during 1987 and 1988. More pod set and less loss to witches' broom and *M. roreri* meant that ripe fruit production in Area "A" was up by at least 50% in 1989 and 1990, relative to 1987.

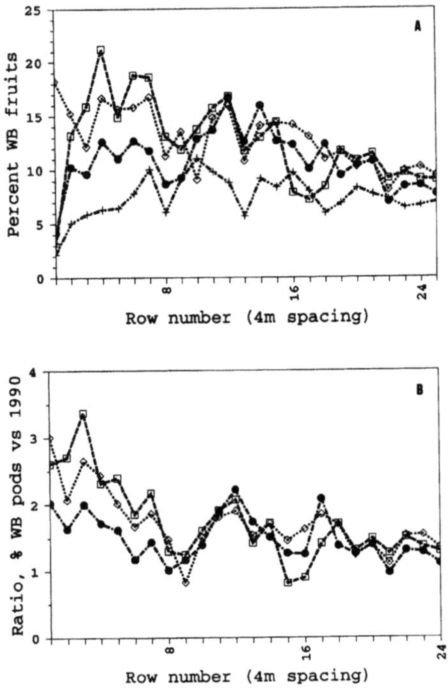

Figure 13.14. (A) Row by row percentage witches' broom pods in Areas "A" and "B" of the Palma Chávez plantation, Pichilingue, Ecuador, 1987–90. (B) Percentage witches' broom pods expressed as a ratio to the 1990 values.
[×, 1986; ◊, 1987; □, 1988; ●, 1989; +, 1990]

San Juan – efficiency of broom removals. Broom removal efficiencies were only ca. 85% in the San Juan and Santa Aurelia plots in 1986, a figure which reflects the difficulty of reaching all parts of the very high canopy, and the large number of brooms present at the beginning. In 1987 the efficiency had increased to 93%, which meant that in January 1988 there were still as many as 8.6 and 18.2 brooms per tree, respectively.

San Juan – infections on shoots and cushions. At the calibration count in 1986, trees in the isolated San Juan plantation had a mean of 278 brooms/tree compared to 578 in Santa Aurelia. Broom counts were down in 1987 relative to 1986 in both sites, perhaps in part because the 1986 removal included brooms from several previous years, while those for 1987 and 1988 were of one year only. The differential between the two sites seen in the calibration counts was maintained but not increased, even though all trees in the San Juan plot had brooms removed, while the Santa Aurelia plot was completely surrounded by heavily infected trees (Figure 13.15). Cushion brooms were much more numerous than those on shoots.

San Juan – witches' broom infection on pods. As in Palma Chávez, because of time limitations there was no specific calibration of pod production and disease incidence. Nevertheless, it was clear from the early harvests which had formed before brooms were removed that the two plantations were very different. San Juan was much more productive, but also lost more to disease, especially to *M. roreri* (Figure 13.15). There was no evidence that the broom removal had reduced pod losses to witches' broom in either San Juan or Santa Aurelia Plot 1, but without calibration or replication, this remains open. What is clear is that there was still considerable pod loss even after cleaning all the trees in San Juan to a higher standard than a farmer would ever be expected to attain in such tall cocoa.

SUMMARY OF RESULTS OF SANITATION

The objectives of the phytosanitation study, to evaluate the efficacy of broom removal and to measure gradients of witches' broom disease in a range of disease pressures, were attained. The presence of environmental gradients in Area "B" complicated evaluation of the results at some sites, but reference to the calibration data allowed detection of broad changes after sanitation.

The response in Area "B" varied considerably among sites and infection courts. In most instances, there was a striking reduction in the mean number of symptoms relative to Area "A"; exceptions were seen for shoots and pods at Balao Chico, and on pods at Táchira. The form of the reduction in "B" also varied. Clear, decreasing gradients away from the non-sanitised area were seen for all three infection courts at Altamira, and for shoots and flower cushions at Ouro Preto. Elsewhere, the falling away of infection with increasing distance was less pronounced or absent. Sometimes, as with pods at Ouro Preto, there was a step down from "A" to "B", but no gradual decrease within "B".

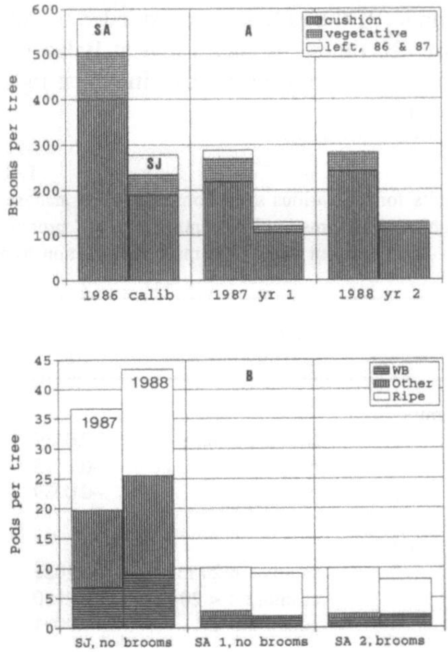

Figure 13.15. Results from the San Juan (SJ) and Santa Aurelia (SA) plantations, Ecuador. (A) Broom counts, 1986–88; (B) Pod production and disease, 1987–88.

The presence of disease gradients implies that while broom removal had a beneficial effect, inoculum from Area "A" did move into Area "B". Background inoculum (from missed brooms within "B" and from distant sources) appeared to play a part too, because in many cases disease did not continue to decrease beyond a certain distance from Area "A".

In general, numbers of symptoms on flower cushions were reduced more readily by sanitation than those on shoots or pods. At a given site, the slope of the curve for cushion symptoms was steeper (Table 13.1), and the levels of "background" infection were reduced more, than for shoot and pod infections. Possible reasons for this might be the greater exposure of shoots to inoculum from afar, and the longer susceptible period for pods relative to individual shoots and flower cushions. The gradients recorded for proportional infection in the seedling trial in a cleared corridor in forest at Rio Negro, Manaus, (Chapter 12) were shallower than those recorded on shoots in the Brazilian sanitation sites. This might be explained in part by the interception of spores by the mature cocoa canopy, though the two situations also differed in many other respects.

The lack of marked gradients at Balao Chico, Ecuador, contrasted sharply with the exponential decreases at Altamira, Brazil, and certain other sites. But at Balao Chico there was a much larger source area than elsewhere, and increasing the area of the source would be expected to flatten disease gradients (Gregory, 1973). As an illustration of the differences in source size, there were approximately five times more diseased trees within 50 m of the midpoint of the "A"/"B" junction at Balao Chico than at Altamira, and ten times more brooms. If the radius were extended to 200 m, the differential in tree numbers increased to nearly 20 times. It seems that at Balao Chico there was sufficient movement of inoculum from "A" and beyond to give infection right through "B", such that gradients were slight or lacking.

TABLE 13.1

Results for the various sites from regression analyses of ln adjusted row mean values for counts of shoot and cushion symptoms, and percentage witches' broom pods in Area "B" versus distance (m) from Area "A" in the treatment year. The multiple infection transformation was applied to percentage pod infection

Site/year	Infection court /distance	Slope (*b*)	Intercept (ln *a*)	R^2	F
Altamira					
	shoots : < 40 m	−0.0489	3.97	0.570	15.90 **
	cushions : < 40 m	−0.1575	4.45	0.633	17.26 **
	pods : all "B"	−0.0089	2.41	0.488	51.49 **
Ouro Preto					
	shoots : < 20 m	−0.1022	4.33	0.817	31.17 **
	cushions : < 20 m	−0.2100	4.11	0.872	47.86 **
	pods : all "B"	0.0004	3.62	0.002	0.06 NS
Manizales					
8/89–2/90	shoots : <52 m	−0.0511	3.19	0.781	42.79 **
8/89–2/90	cushions : all "B"	−0.0136	3.51	0.594	49.79 **
12/88–6/90	pods : all "B"	−0.0060	2.95	0.121	6.63 *
Guamal					
8/89–2/90	shoots : all "B"	−0.0042	2.47	0.042	1.50 NS
8/89–2/90	cushions : all "B"	−0.0059	4.42	0.089	3.34 NS
Balao Chico					
	shoots : <80 m	−0.0037	2.15	0.023	0.62 NS
	cushions : <80 m	−0.0073	4.17	0.149	4.57 *
	pods : all "B"	+0.0021	0.0015	0.048	2.00 NS
Táchira					
	shoots : <105 m	−0.0174	3.22	0.440	7.88 *
	pods : all "B"	−0.0009	3.94	0.310	6.14 *

At the more isolated Palma Chávez site in Ecuador, the source was similar in size to those at the protocol sites which had gradients. Gradients were recorded on shoots and flower cushions at Palma Chávez, but not on pods. When the entire plantation was cleared of brooms very carefully, there was still sufficient background inoculum to give rise to an 8% pod loss to witches' broom, and an average of 40 brooms per tree. Presumably, inoculum coming from other plantations in the vicinity was largely responsible for this infection.

The amount of incoming inoculum no doubt differed among the protocol sites, because disease severity and the amount of non–sanitised cocoa around the sites varied. In Brazil, for example, even though the two sites had a similar degree of isolation from the nearest neighbouring cocoa, farms in Ouro Preto tended to be contiguous rather than widely scattered as in Altamira, where large sugarcane plantations and primary forest occurred between. Moreover, as a result of low cocoa prices, Ouro Preto farmers had neglected phytosanitation measures for two seasons and this contributed to a build–up of inoculum in the region, which might explain, at least in part, the higher background infection encountered there. Another factor to be considered is that despite similar broom removal efficiencies in the two sites, the initially higher broom load in Ouro Preto resulted in more brooms being left per tree than in Altamira.

In the Colombian sites, the absence of adequate calibration data and the sub–optimal timing of the broom removals made the identification of gradients less certain, but shoot, cushion and pod symptoms in Manizales did seem to decrease for some distance from "A". The Guamal site showed no gradients within Area "B", but they may have been obscured by infection from brooms left in the area after the rather inefficient removals.

In Venezuela, results from the comparative epidemiology study became available while the phytosanitation experiment was in progress. They showed that broom removal in August, as practised in the sanitation trial, was rather late. This may have been the main reason for the limited effect of phytosanitation on the number of pods damaged by witches' broom, though there was an appreciable reduction in cushion and shoot disease. Broom removal, when carried out in March/April, is a valuable practice in the conditions of El Piñal, Táchira State, and has enabled the few cocoa producers there to maintain profitable levels of production. Appropriately–timed sanitation has also been shown to be effective in reducing pod losses in commercial farms in the Brazilian Amazon and in Colombia.

The beneficial effects of sanitation were not limited solely to reduction in disease. There was also an increase in pod production relative to that in Area "A" in most of the sites between the calibration and the treatment years (Figure 13.16). This was usually reflected in increased yields of ripe pods and dry cocoa. Lower productivity in "A" presumably was due to the reduced ability of symptom–carrying flower cushions to bear fruit, to energy and nutrient wastage in the production of diseased tissue, and to loss of photosynthetic capacity in trees with severe canopy infection.

The attempts in Ecuador to reduce pod loss in 8 m tall farmers' cocoa failed, as had Thorold's efforts in similar cocoa in Trinidad (1943), though in his trial brooms were left on the ground beneath the trees after removal. In Ecuador, brooms were carried away and destroyed, so the cause of the pod loss there would have been the brooms unavoidably left in the inaccessible canopy, and background inoculum from afar. It was not possible to estimate the relative contributions of these two potential sources.

The IWBP sanitation studies relate to relatively few sites and to a single year, but they have advanced knowledge on disease gradients of witches' broom considerably. They have revealed a wide range in the response to broom removal in terms of the reduction in disease achieved and the form of the disease gradient. This is not too surprising perhaps as the sites differed in many respects, including initial disease severity, the efficiency of broom removal

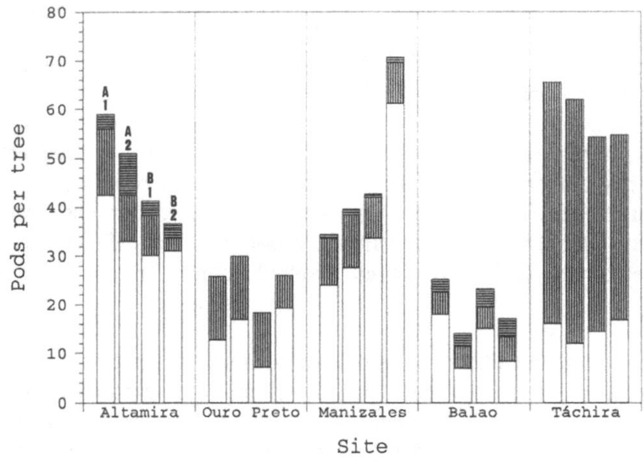

Figure 13.16. Production of healthy and diseased pods in areas "A" and "B" in the calibration period (Year 1) and after sanitation (Year 2) at the various comparative phytosanitation sites. Open bars, ripe pods; vertical stripes, witches' broom pods; horizontal stripes, other diseases.

in Area "B" and adjacent cocoa, the shape and extent of Area "A", and the proximity and extent of other cocoa with witches' broom. They no doubt differed also in poorly or undocumented characteristics which would affect dispersal such as the direction, scale and timing of air movements through "A", "B" and nearby cocoa. The sites were too few in number to permit multivariate analysis of such parameters as gradient steepness in relation to known site characteristics, and further studies will be necessary at more locations if the factors controlling the form of gradients are to be isolated and quantified. Only then might it be possible to predict with some certainty the outcome of sanitation for a specific situation. Nevertheless, in most circumstances the prognosis for effective sanitation appears to be good (see "implications for disease management" below).

The demonstration of dispersal through the cocoa plantation and from one farm to another raises difficult questions for the researcher, such as "how much isolation between farms is necessary to avoid problems from witches' broom disease?" In his review of disease gradients and control measures, Thresh (1976) pointed out that the minimum effective isolation distance is not constant for a particular disease. He believed that this concept is seldom understood by administrators responsible for official policy, or by growers and extension officers; it is often naively assumed that spread does not occur beyond a set distance that may be selected quite arbitrarily. The results of the IWBP comparative phytosanitation studies confirm that there is no single gradient to describe witches' broom dispersal. Moreover, gradients for the three infection courts differ at a given site, and a separation distance which would give an acceptably low rate of flower cushion infection for example, might not be sufficient for disease on shoots and pods. Refinement of our understanding of witches' broom dispersal depends on further field experimentation, be it in the form of carefully executed demonstration plots, or specifically designed research studies.

IMPLICATIONS FOR DISEASE MANAGEMENT

1. Careful phytosanitation reduces disease and increases yield in many situations, even when there is infected cocoa nearby.

2. The demonstration of disease gradients shows that there is a "neighbour" effect, that uncleaned cocoa does act as a source of witches' broom infection for adjacent cocoa, and can therefore reduce the effectiveness of sanitation treatments.

3. The degree of interference caused by an uncleaned area is difficult to predict because it depends on many site-specific factors which include; disease severity in the region, the relative position, shape and size of the cleaned and uncleaned areas in question, topography, prevailing air movements, and the efficiency of broom removal.

4. In regions with large expanses of heavily infected cocoa which lack sanitation interference is likely to be high, especially if the cleaned area is just a small pocket amidst a vast area of cocoa infected with witches' broom. In this case, the beneficial effects of broom removal might be restricted to an increase in overall pod production, which may or may not be reflected in more healthy pods. For sanitation to be effective in reducing disease in contiguous cocoa, it has to be applied generally over the whole region.

5. Where cocoa farms are scattered, interference will be less, but may still cause difficulties in high risk regions, both from direct losses and because of the need for repeated sanitation year after year as a result of reinfection of the canopy and flower cushions. In regions of lower risk, sanitation is likely to be successful and the prevention of build-up feasible, provided that trees are of a manageable height.

6. Tall, unpruned cocoa is practically impossible to clean thoroughly.

7. Appropriate timing of broom removal is essential if phytosanitation is to be successful.

8. The "neighbour" effect may cause interplot interference in trials and demonstration plots, which might also be swamped by inoculum from adjacent or nearby uncleaned cocoa. Guard areas would help, especially for studies on shoot and flower cushion disease, but if guard areas are too small or absent, the tendency will be to undervalue successful treatments or cultivars, or miss their usefulness completely. Vanderplank (1963) gave an assessment of

this general problem in field experimentation.

9. A better understanding of the reasons for the environmental gradients in productivity and symptoms would provide clues to improved all-round management.

Chapter 14

WITCHES' BROOM IN BAHIA, BRAZIL

H.M. Rocha, R.A.C. Miranda, R.B. Sgrillo & R.A. Setubal

INTRODUCTION

Approximately one–fifth of the world's cocoa beans are produced in Brazil, of which nearly 85% originates from around 600,000 ha of the crop concentrated in the southern part of the State of Bahia. The remaining 15% comes from four states in the Amazon Basin (Rondônia, Pará, Amazonas and Acre) and the states of Espírito Santo, São Paulo and Mato Grosso. Two major pests (*sensu latu*) cause serious losses of cocoa in Brazil, of which the most serious in overall economic terms, *Phytophthora* pod rot, is found throughout the country. Until the recent outbreak in Bahia, the other major pest, witches' broom disease, was only found west of a line from Maranhão state to Mato Grosso. This chapter presents the history of the outbreak in Bahia, the initial response to try to eradicate it, followed by attempts at containment, and finally the efforts to manage the disease. The phenology of cocoa growth is considered using data collected according to the IWBP experimental protocol, before the first record of witches' broom in Bahia in the central part of the cocoa zone. These data are compared with data from other IWBP sites where the disease was present, and predictions are made of the dynamics of the disease as it becomes further established in Bahia.

WITCHES' BROOM OUTBREAK IN BAHIA

Possible Methods of Introduction

Until May 1989 when the first disease outbreak in the region was confirmed, a phytosanitary cordon had been organised around Bahia and Espírito Santo to try to exclude witches' broom from those states. Inspection points were established at the main airports in the Amazon region, and no plant material was allowed onto flights to other regions. Further inspection points were operated at airports in Salvador and Ilheus in Bahia. Movement of plant material by road was not screened as thoroughly, partly because road connections from the Amazon region to the south and east were only improved sufficiently to allow significant traffic in the late 1980s.

Based on existing knowledge of the biology of *Crinipellis perniciosa*, it is extremely

Disease Management in Cocoa: Comparative epidemiology of witches' broom. Edited by S.A. Rudgard, A.C. Maddison and T. Andebrhan. Published in 1993 by Chapman & Hall, London. ISBN 0 412 58190 6

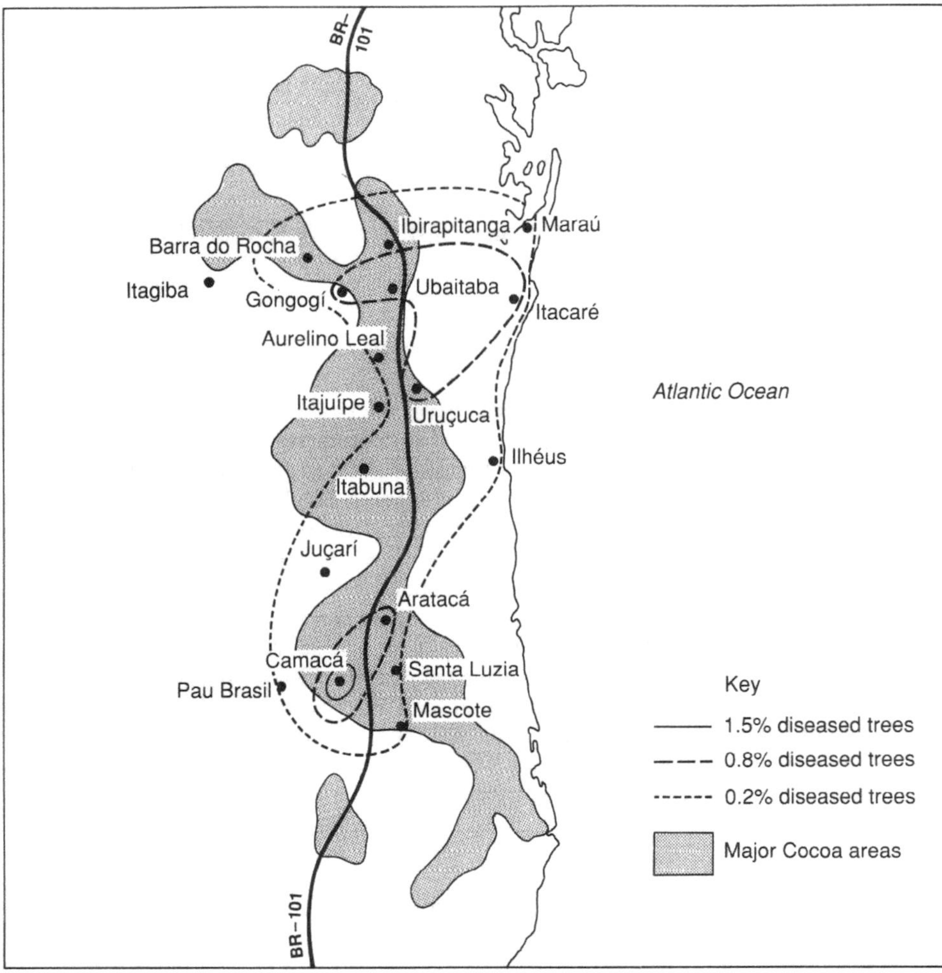

Figure 14.1. Distribution of witches' broom disease in Bahia, Brazil in 1991 (calculated as % diseased trees).

unlikely that witches' broom was introduced into Bahia by wind-blown spores. The prevailing winds in southern Bahia are the south-easterly trade winds, which would stop long-distance dispersal from the Amazon towards the coast. Basidiospores are extremely vulnerable to desiccation and ultra-violet light (Baker & Crowdy, 1943), and it is extremely unlikely that they could have remained viable after traversing nearly 2,000 km of semi-arid area that separates the cocoa region of Bahia from the nearest affected area in the Amazon.

The most likely method of introduction would have involved carriage of the fungus overland in one of two forms. The saprophytic phase of the fungus could have been introduced as mycelium or chlamydospores in fragments of or entire necrotic infected tissues. The fungus could also have been introduced in its parasitic form in infected living cocoa tissue, transported in seedlings or in seeds in unbroken diseased pods. Large areas of cocoa were established in the Brazilian Amazon after 1975, with 90% found in Rondonia and Pará states. After construction of the BR-364 road to Rondonia, some traders started to take cocoa beans to Bahia for export. Research has shown that there is no danger of transmission in properly fermented and dried beans (Almeida, 1988), but not all seedlings and seeds were intercepted at the phytosanitary cordon. It is a matter of speculation as to whether the introduction of witches' broom into Bahia was accidental or intentional.

First Reported Outbreaks

The disease was first identified in May 1989 in the county of Uruçuca in the centre of the main cocoa belt in Bahia (Figure 14.1), where an old commercial plantation of about 1.2 ha was found showing various vegetative and cushion symptoms of the disease (Pereira *et al.*, 1989). Remedial action was taken immediately (see below). In late October of the same year, approximately 60 days after the eradication work had finished at the first farm at Uruçuca, another 20 infected plants were found close by the cleared area of the first focus. Then in November, several disease foci were found in the county of Camacã, located 120 km south of Uruçuca. The two major foci at Uruçuca and Camacã were close to the BR101 main road that runs from Rio de Janeiro to Salvador and carries the heavy traffic through the region. This may only be a coincidence, with no relation to the first introduction of the disease. New outbreaks of the disease have since been reported from farms in many counties spread throughout the main cocoa belt of southern Bahia.

It is interesting to note that, although most classic symptoms of witches' broom disease were found, no infected pods had been observed up to the middle of 1991. The first records of diseased pods occurred late in that year, some 2.5 years after the first record of witches' broom in the region.

Disease Spread

After the detection of the first foci in Uruçuca and Camacã, new reports of the disease showed no particular pattern. Infected farms were found throughout the major cocoa areas, with the exception of the most northerly around Gandu and the most southerly around Belmonte. CEPLAC records in mid-1991 showed that the disease had been identified on 994 farms, with over 400,000

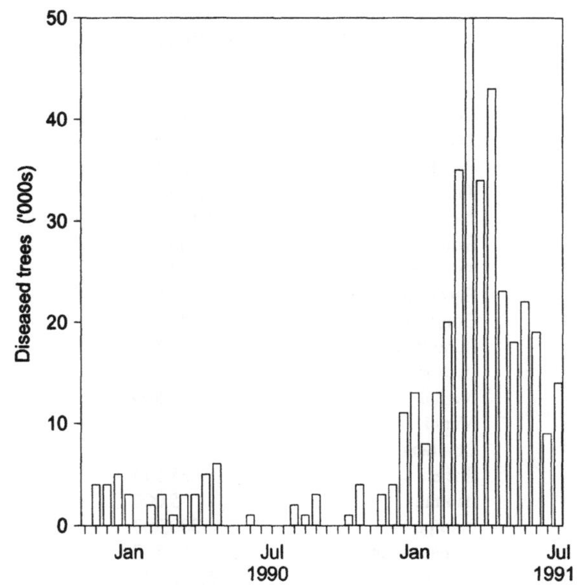

Figure 14.2. Variation in the number of trees newly infected with witches' broom between November 1990 and July 1991 in Bahia, Brazil.

infected plants scattered in approximately 50,000 ha located in 19 counties (Table 14.1). Witches' broom incidence was greatest in Camacã, where 3.4% of trees inspected had been found to be diseased. A map of disease incidence in 1991 derived from Table 14.1 indicates that the two major concentrations of the disease can be distinguished (Figure 14.1).

The most serious outbreak around Camacã at the southern end of the affected region contained over 60% of all diseased trees recorded in Bahia. The neighbouring Aratacá county also had high incidence of 1.65% diseased trees. The more northerly Itacaré and Uruçuca counties were the only other areas with more than 1% disease, and the above four counties accounted for over 90% of all diseased trees found up to 1991.

It is unclear whether witches' broom disease arose at one initial location in Bahia from which it then spread, or whether primary outbreaks occurred simultaneously in more than one place (e.g. Camacã and Uruçuca). The prevailing wind direction in the region is from the south to the north. The disease may have spread in a northward direction from a single primary focus in Camacã, some 20, 120 and 150 km to the three major foci mentioned above. Secondary spread may then have occurred from all four sites giving the pattern in Figure 14.1. The primary dispersal must have occurred before May 1989, but CEPLAC's data are inadequate to define how long the disease had been in Bahia before it was first identified. If there was only one disease cycle per year as in the Amazon, the epidemic would need at least one year to build up at each site, and it is probable that the disease had been in the region at least since 1987 and perhaps even earlier.

TABLE 14.1
Numbers of trees inspected and infected trees in Bahia, Brazil up to July 1991

County	Trees inspected	Trees infected	% infected
Camacã	7,519,628	257,194	3.4
Mascote	300	3	1.0
Pau Brasil	884,142	6,367	0.7
Aratacá	2,719,485	40,888	1.5
Juçarí	1,125,241	4,832	0.4
Santa Luzia	163,078	326	0.2
Aurelino Leal	2,042,078	9,174	0.4
Barra do Rocha	97,085	154	0.2
Gongogí	142,288	1,276	0.9
Ibirapitanga	1,176,574	3,244	0.3
Itacaré	1,262,720	20,834	1.6
Itagibá	78,500	191	0.2
Maraú	277,470	751	0.3
Ubaitaba	85,100	697	0.8
Ilhéus	1,638,540	5,626	0.3
Itabuna	5,000	27	0.5
Itajuípe	380,300	330	0.1
Lomanto Junior	66,847	16	0.1
Uruçuca	6,940,633	72,378	1.1
Total	26,605,862	424,308	1.6

Most new identifications of infected trees in 1990 and 1991 occurred between January and May, and very few during the cool period of May to August (Figure 14.2). Although the data are not conclusive, the single annual increase in the number of infected trees supported the hypothesis that witches' broom maintained one annual cycle in Bahia. The timing of disease processes is discussed later in the chapter.

Containment of Outbreaks

An attempt to eradicate the disease at the sites confirmed in 1989 was coordinated by the offical government body responsible for cocoa in Brazil, CEPLAC. At the infected farms in Uruçuca, the entire affected cocoa plantation and surrounding forest for 350 m in all directions were razed and burnt, representing a total area of about 100 ha in each case. These methods were abandoned after 1989, when the number of reported outbreaks became too numerous to deal with in this fashion. It also had become clear that quick eradication from the region was unlikely, and that a sustained campaign of containment was the more logical strategy.

CEPLAC then implemented a campaign to promote phytosanitation involving the participation of researchers, extensionists, farm associations and the community in general. Practices for removal of infected tissues, pruning of trees and monitoring of reinfection were established. The campaign had several aims apart from the implementation of phytosanitation, as cocoa growers had to be educated about the disease, trained how to manage the crop, and motivated to do so thoroughly. The intensive training programme for labourers and farm managers involved sessions at different localities in the region, in which over 60,000 persons were trained.

CEPLAC itself coordinated a systematic inspection programme for identification of infected farms throughout the region. Pruning teams of labourers then worked through such farms, carefully removing every diseased plant part (vegetative and flower brooms) and fructifying material from the trees and ground under them. The removed material was taken out of the plantations and either burned or buried. Up to mid-1991, a total 965 farms had been treated. Of the 424,000 infected trees identified in Table 14.1, almost 34,000 (8%) severely infected plants were cut down completely and about 220,000 had been pruned by July 1991.

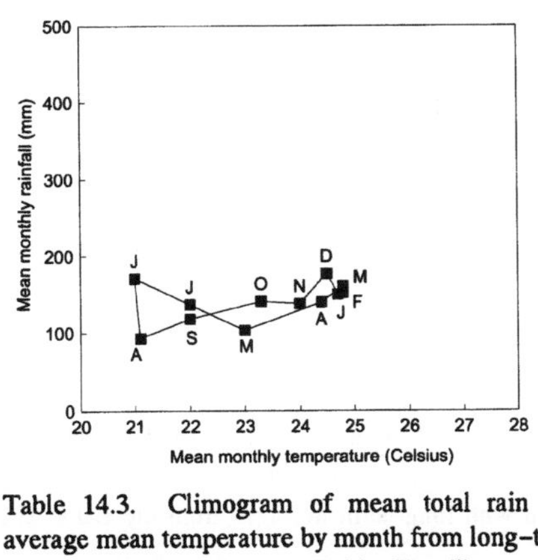

Table 14.3. Climogram of mean total rain and average mean temperature by month from long-term climate records at Itabuna, Bahia (Brazil).

COCOA PHENOLOGY AND CLIMATE AT ITABUNA, BAHIA

The IWBP programme for comparative epidemiology studies incorporated a site in Bahia at CEPEC at Itabuna. Witches' broom disease was not present at the site throughout the period of data recording, which began in 1986 and finished in 1988, and had still not been recorded at CEPEC by 1992. The disease outbreak in Bahia was confirmed as the analysis of the IWBP epidemiology data was in progress, increasing the importance of the study at Itabuna.

The objective was to gather data on host growth and climate which could be compared with sites where the disease was present. It was hoped that predictions of the possible behaviour of the disease at that site could be made using those data. The methodology followed the protocol for the other sites, with the exception of disease incidence and basidiocarp production.

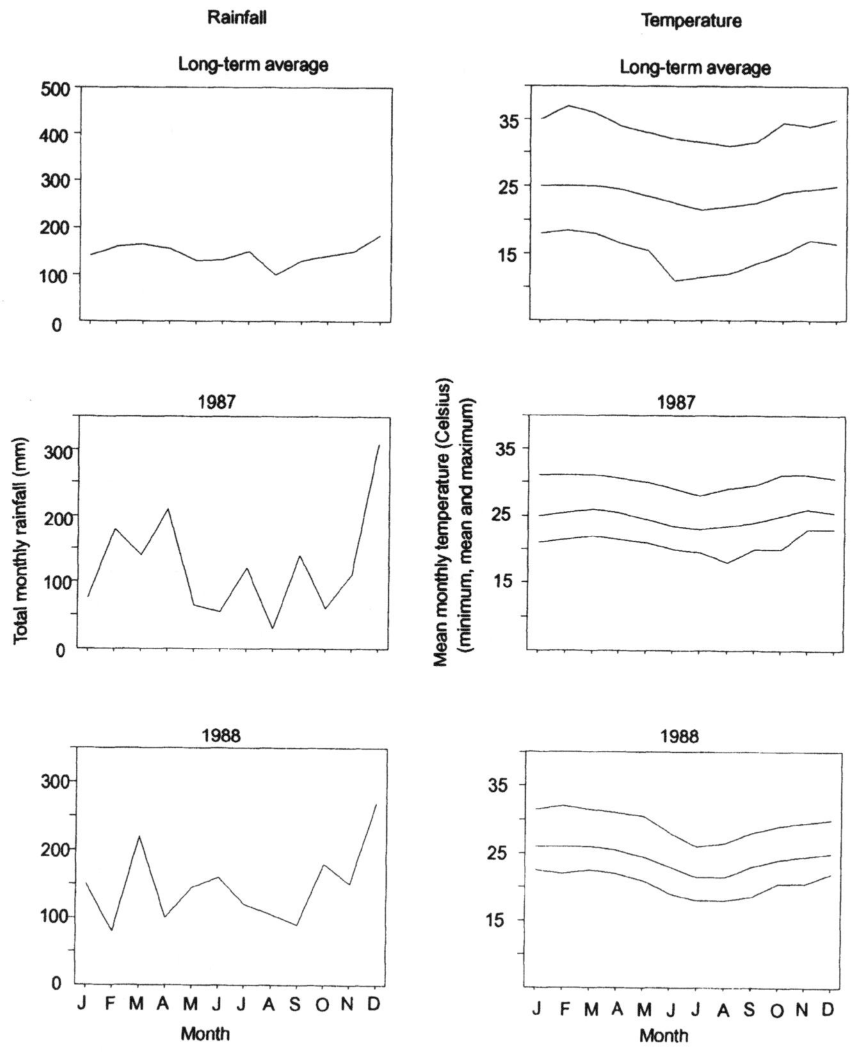

Figure 14.4. Rainfall and temperature data compared with long-term averages from the Bahia Site, Brazil.

Climate

The cocoa growing area of Bahia is classified climatically as being humid tropical, with a total rainfall of between 1,400 mm and 2,000 mm. Itabuna is part of the "Almada" agrosystem, which is the most traditional cocoa growing area of Bahia (Sá *et al.*, 1982). The site climograms derived in Chapter 11, showing long–term monthly averages of rain plotted against monthly averages of temperature, were compared with a climogram for Itabuna (Figure 14.3). Itabuna differed from all the IWBP sites except Manizales by the absence of clearly defined annual wet and dry periods. All months had long–term average rainfall of less than 200 mm and, except for August, more than 100 mm. The mean annual temperature was about 23°C, which was appreciably lower than most of the other sites studied in the IWBP. Four months (July–September inclusive) had average temperatures of less than 22°C, which was the minimum monthly average at the other IWBP sites. In summary, long–term averages from Itabuna showed the particular characteristics of even rainfall distribution, and a large temperature range (4°C) with a distinct cool season.

Rainfall and temperature for the two years of the IWBP experiment are shown with the long–term averages in Figure 14.4. The monthly averages for the long–term data all had large standard deviations, and rainfall pattern varied greatly between individual years. Year 1 (1987) and year 2 (1988) at Itabuna had 5 and 2 months, respectively, with less than 100 mm of rain, and 2 months each with more than 200 mm. December had the greatest rainfall in both years, and yet January (year 1) and February (year 2) had the least. There was only one month in the two years with less than 50 mm (August, 1987). The long–term temperature profile with a cooler season from July to September was maintained during the two years of the IWBP. Maximum and minimum temperatures remained below 33°C and above 17°C, respectively, which was an appreciably smaller amplitude than normal.

Cocoa Phenology

The phenology of cocoa in Bahia has been intensively studied (Alvim *et al.*, 1974; Almeida *et al.*, 1987), but further data on growth and fruiting were recorded at the IWBP site in Itabuna in order to make a direct comparison with the other IWBP sites. Patterns of shoot growth were similar in the two years of the IWBP, with the most intense flush around weeks 34 and 36 (September) and 2–3 less intense flushes between weeks 46 and 12 (December–March) (Figures 14.5 and 14.6). There was an extended period without substantial shoot growth from week 12 to week 32 of both years, which spanned the cooler season.

There was a clear single annual cycle in the number of flower cushions active, with maximum intensity of activity between weeks 50 and 12 (December–March) (Figures 14.5 and 14.6). Flowering almost ceased between weeks 16 and 40, which was the cooler season. There were two distinct periods of pod harvest in the year, with the main crop between weeks 38 and 49 (October–November), and a secondary harvest season between weeks 14 and 22 (April–May). The two periods of pod production overlapped, so that the secondary crop was starting to develop when the main crop was being harvested. Flowering activity began to increase during the main pod harvest, and began to decrease during the secondary harvest.

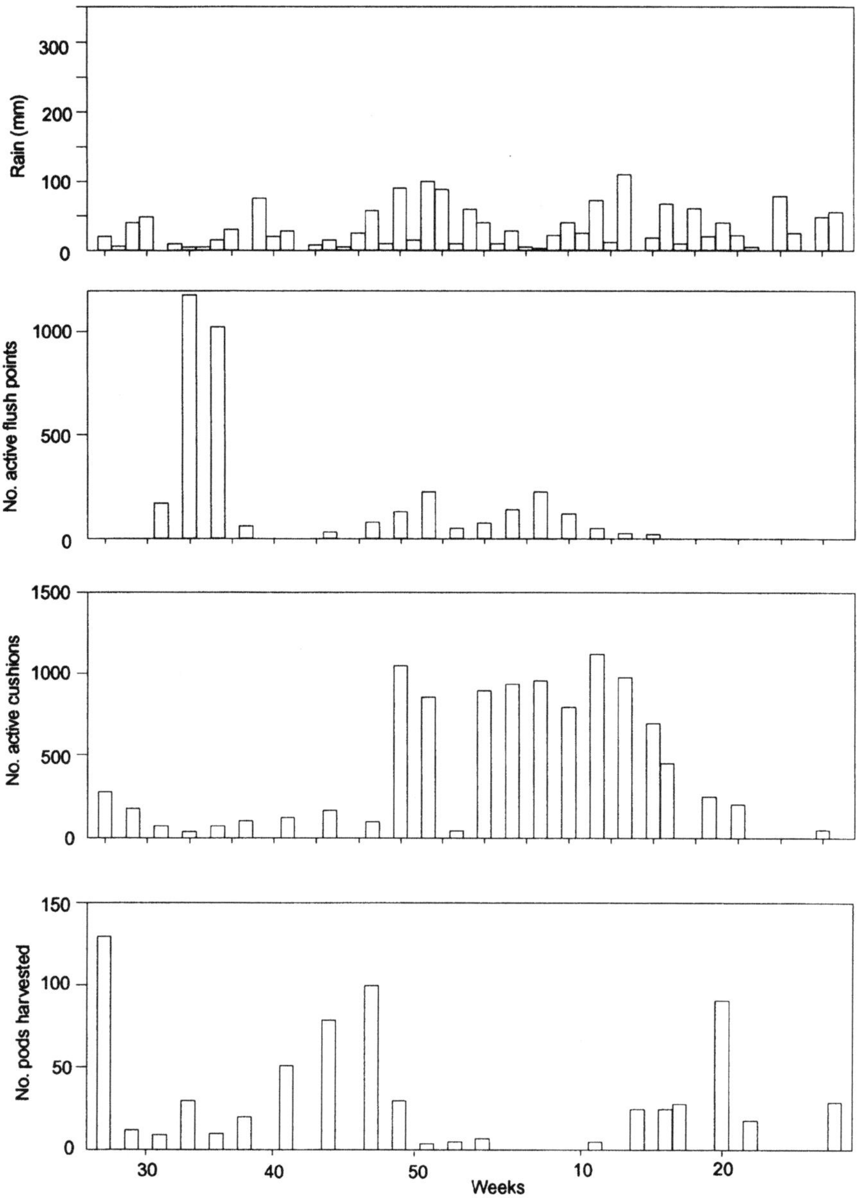

Figure 14.5. Rainfall, vegetative flushing, flower cushion activity and fruiting in the Bahia Site, Brazil in 1986/7.

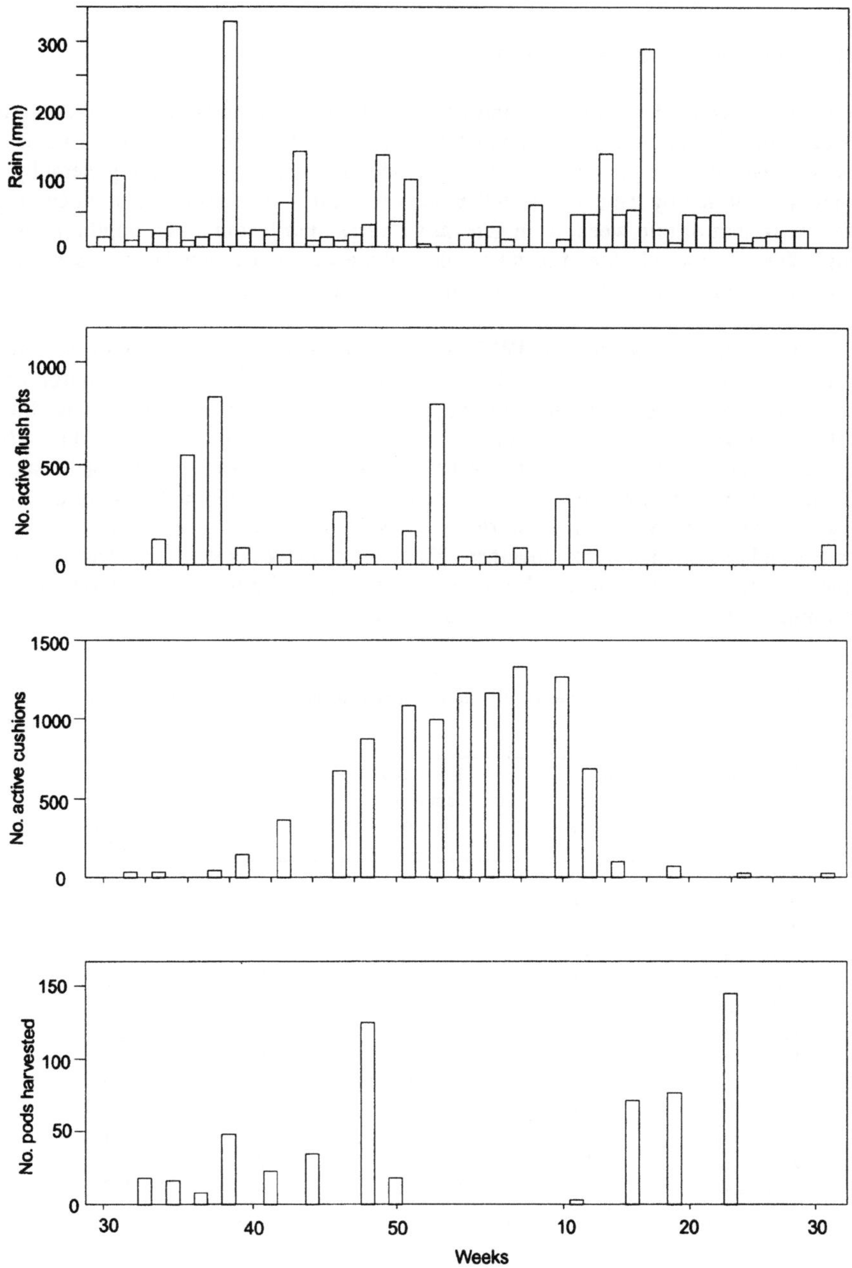

Figure 14.6. Rainfall, vegetative flushing, flower cushion activity and fruiting in the Bahia Site, Brazil in 1987/8.

PREDICTION OF WITCHES' BROOM EPIDEMICS

IWBP Site : Inoculum Basidiocarp Production

The comparative epidemiological analysis in Chapter 11 presented data on the relationships between climate and basidiocarp production in nine IWBP sites. The main conclusion from those sites was that the period of basidocarp production was determined in general by climate. This was supported by the conclusion that no basidiocarps were normally produced in periods with less 10 mm of rain/week, or in general with less than 100 mm per month. However, the relationship did not hold for Manizales, where there was no extended dry season, but several months per year with 10–100 mm of rain in which basidiocarps were produced.

The climate data from the IWBP site at Itabuna were examined with the intention of predicting when basidiocarps would be produced on brooms placed in that environment. The absence of an extended dry season at Itabuna make it likely that, as in Manizales, drier weeks would have less of an inhibitory effect on basidiocarp production than they did at IWBP sites in general. The rainfall data for the IWBP site at Itabuna showed a total of 36 weeks with less than 10 mm of rain over the whole of years 1 and 2 combined. Many of these drier weeks occurred singly, but there were six cases of two weeks together, two cases of three together and one of four weeks together. There were 33 weeks with less than 10 mm of rain in the two years at Manizales. The number of weeks at the two sites without any rain was also similar, at 11 and 13 for Itabuna and Manizales, respectively.

TABLE 14.2
Rainfall characteristics at the IWBP site at Itabuna, Bahia (Brazil) in 1987 and 1988

Time period (h)	No. of rain events	Mean rain amount (mm)	Mean duration (h)	Mean intensity (mm/h)
01.00–02.59	89	5.9	2.3	1.7
03.00–04.59	3	82.3	2.1	0.8
05.00–06.59	96	3.2	2.1	1.5
07.00–08.59	79	3.7	1.8	1.7
09.00–10.59	109	3.0	1.7	1.5
11.00–12.59	73	5.0	1.9	2.2
13.00–14.59	66	2.8	1.9	1.2
15.00–16.59	36	7.4	2.3	2.0
17.00–18.59	33	6.6	2.9	2.1
19.00–20.59	33	6.8	2.7	1.4
21.00–22.59	42	6.0	3.2	1.3
23.00–24.59	68	3.1	2.4	0.7

Basidiocarps were observed throughout almost the entire year at Manizales, although they may not have been turgid and actively releasing basidiospores in weeks without any rain. Considering only the rainfall pattern, it is likely that basidiocarps would be formed throughout most of the year in Itabuna in years similar to 1987 and 1988. However, extended dry seasons have occurred in Bahia, and such an event would clearly inhibit basidiocarp formation.

The amount of basidiocarps produced in the IWBP sites (Chapter 11) was found to be promoted by good daily drying conditions in the rainy season, such as low daytime humidity and rainfall at night or in the early morning. Most rainfall at Itabuna occurred between 05.00 and 11.00 h (Table 14.2), and the daytime humidity was low compared with the other IWBP sites. It is possible that basidiocarp production would be favoured by the Itabuna climate, as indeed it was at Manizales, although the scale of such production need not directly influence disease.

IWBP Site : Occurrence of Witches' Broom Disease

The likely timing of disease at Itabuna could theoretically be derived relative to the predicted period of basidiocarp production above and the phenology of growth. However, it is important to note that quantitative prediction is not possible with the data available, and prediction is restricted to timing of disease.

It was predicted earlier on the basis of the climate data that basidiocarp production would have occurred through most of the year in the IWBP site at Itabuna. Data from Manizales were used as a basis for prediction of basidiocarp formation at Itabuna due to the similarity of the climate with that of Bahia, and were also useful in considering disease. The almost continuous production of basidiocarps observed in Manizales was associated with continuous incidence of disease symptoms, especially for flower cushions which were active for most of the year. However, there were related peaks of disease incidence on shoots, cushions and fruit that indicated that disease potential was not equivalent throughout the year. It was not known whether infection potential had remained similar. This situation would probably occur in Bahia, and presence of basidiocarps throughout most of the year in Itabuna might be associated with one or more peaks of disease incidence.

The phenology of cocoa at Itabuna would determine the availability of infection courts, and thence the disease potential. An important restriction on the development of disease in 1987 and 1988 would have been the almost complete lack of active shoot growth from week 14 to week 30, and lack of flower cushion activity from week 20 to week 40 (Figures 14.5 and 14.6). These periods of inactivity defined the times of least exposure of susceptible vegetative and floral tissue. The greatest volume of susceptible vegetative tissue occurred during the main shoot flush in September, and peak flowering in January to April. This would have been the most likely time for infection to have occurred in those years, had basidiocarps been present. It was not clear whether the phenology of growth observed in 1987 and 1988 was truly representative, but the similarity of the two IWBP years was striking.

Bahia State : Occurrence of Witches' Broom

Patterns of flushing, flowering and fruiting were known to vary slightly through the different agroclimatic zones of the region (Carzola *et al.*, 1989). The relative timing of meristematic activity would be of great importance to the development of the epidemic in the various zones. Such variations would qualify an evaluation of the potential incidence of *C. perniciosa* elsewhere in Bahia using only the IWBP data from Itabuna.

It was interesting to note that most new identifications of infected trees in Bahia by CEPLAC extension staff, as mentioned above, occurred between January and May. New sightings would probably have been linked to times of peak production of symptoms, and it might be inferred that peak disease incidence occurred in the first four months of the year. Once allowances were made for incubation periods, it could be speculated that the main period of infection spanned at least the months from November to March. It has already been stated that peak cushion activity in the IWBP site at Itabuna occurred during those months, together with appreciable shoot growth. Although it was not possible to derive firm relationships from these extrapolations, the coincidence was interesting and merits further investigation now that the disease is in the region.

Concerning pods, it was also interesting to note that the main crop harvested in October/November would have been susceptible to infection between May and August, in contrast to the main period of susceptibility of vegetative and flower tissue from November to March. The factors determining the intensity of pod infection by witches' broom are not entirely clear from the IWBP studies.

Itabuna : Timing of Disease Management Practices

It is also necessary to consider application of the main method for management of witches' broom, phytosanitation, were the disease to become established in a plantation with growth phenology similar to that of the IWBP site. In the Amazon region, the climate is divided into well-defined wet and dry seasons, and witches' broom disease shows a clear annual cycle determined by the seasons. The latter influences the timing of phytosanitation, which is completed in the dry months, August and September. The Bahian climate is not so clearly defined, and data on the disease cycle are scarce.

It has already been stated in earlier chapters that broom removal is the key practice in sanitation, and correct timing is essential. Circumstantial evidence from other parts of Bahia indicate that broom formation occurred between January and May. The main structural pruning to maintain canopy architecture is normally carried out in the period between January to March, which would be in the middle of the apparent peak of symptom formation. Removal of brooms at that time would miss those that form late in the season. The clash of phenology and ideal sanitation practices is difficult to resolve, especially related to demands on labour for harvesting and processing beans.

The middle of the period without shoot growth and cushion activity, that is to say June, would be a possible choice. The broom removal would have to be timed to follow immediately after the secondary pod harvest in April–May, admittedly when there would be another demand on labour for processing beans. Sanitation could not be left to later in the year, as it would be inadvisable to prune the canopy after the main pod crop had set in July and August. Consideration of this problem will be greatly facilitated by more data on the timing and nature of the disease cycle in Bahia.

Chapter 15

DISEASE MANAGEMENT : RECOMMENDATIONS

S.A. Rudgard, T. Andebrhan, A.C. Maddison & R.A. Schmidt

INTRODUCTION

As stated in Chapter 2, potential methods for managing witches' broom disease fall mainly into three groups, sanitation, spraying with chemicals, and use of host resistance. Cocoa farmers in some of the regions and countries participating in the IWBP had already been faced with the disease for many years before the project started, and some had been applying management techniques. It would be worth briefly summarizing the situation of disease management at the start of the project. In Brazil and Colombia, action was being taken to manage witches' broom by phytosanitation in areas of hybrid cocoa affected by the disease, but only when the economic climate was favourable. In Ecuador, various management packages had been developed over more than 50 years of research, but such methods were not adopted in the traditional cocoa that made up the bulk of the crop in that country. The disease was a localized problem in Grenada and Trinidad, and very little action was generally taken by farmers. In Venezuela, witches' broom was only a serious problem in one region of the country, where it was a relatively recent introduction. The general response of the farmers there had been to replace cocoa with other crops.

The IWBP was formulated on the basis that a greater understanding of the interactions between weather, inoculum and host phenology would provide an improved basis for the selection or formulation of the most suitable techniques for management of witches' broom disease in a range of cocoa-growing environments. Chapters 5–13 presented summaries of the results obtained by the IWBP and discussed the conclusions relative to disease management. Authors of the individual country chapters (Chapters 5–10), and the two comparative chapters (11 and 13), proposed phytosanitation as the most immediately available method of managing the disease, although there were reservations about the economic viability and effectiveness of the treatment in certain situations.

The techniques and guidelines for the application of phytosanitation are dealt with in this chapter. The other options for the management of the disease were generally agreed to be medium to long-term options requiring varying amounts of development, and some aspects of phytosanitation were considered to require further research. These research areas and the future prospects for the various options are dealt with in the next chapter (Chapter 16).

Disease Management in Cocoa: Comparative epidemiology of witches' broom. Edited by S.A. Rudgard, A.C. Maddison and T. Andebrhan. Published in 1993 by Chapman & Hall, London. ISBN 0 412 58190 6

The scientific data both of the IWBP and of previous workers provide the base material upon which the proposals in this chapter are built. The conclusions of the IWBP studies and the elements of past experience have been combined into a set of strategies for the management of witches' broom disease. The reports from each country (Chapters 5–10) also included a brief discussion of the agronomy of the cocoa crop in each region/country. Plans for disease management need to take into account features of the crop not directly related to pathology, and conversely recommendations for agronomic practices may have an impact on the specific disease–related practices. Consideration is given to the relative impact of management practices on the economics of cocoa production. The recommendations are presented separately for new and existing plantings, and the rehabilitation of less–productive diseased plantations.

NEW PLANTINGS

A farmer establishing a new plantation of cocoa in an area of high witches' broom incidence has more options for managing disease immediately available to him than would be available to the owner of an existing mature plantation. He can make decisions to adopt practices at the outset which will influence the potential production of the plantation.

Planting Material

The main emphasis should be placed on the selection of resistant material, where it is available. There is an urgent need to stimulate selection and breeding to provide better materials. Strategies for such a programme need to be long–term, and are considered in the next chapter. The strategy is mentioned here since the selection of appropriate material can profoundly affect the success of phytosanitation.

Tree Architecture

Good management in young plantations is critical, particularly with regard to control of tree height and form through structural and maintenance pruning to keep canopy height at less than 3.5–4 m in height. Farmers should treat each tree individually, aiming to establish its architecture for maximum ease of harvesting and broom removal. The tree should be restricted to the first jourquette and higher storeys eliminated by regular maintenance pruning to remove basal and canopy chupons. A good tree form may be more difficult to maintain with plantings at densities much in excess of 1,200 trees/ha, although close spacing does allow eradication of heavily infected trees without creating problems of persistent gaps in the canopy.

Phytosanitation in New Plantings

The rate at which witches' broom builds up in a new planting will depend on the amount of background inoculum arriving and the efficiency of any phytosanitation implemented. In isolated plantings, especially in low–risk areas, it would be possible to maintain disease at insignificant levels provided sanitation is started early, and practised conscientiously every year. In high risk areas, background inoculum alone may cause appreciable pod losses however thoroughly the planting itself is cleared of brooms.

TABLE 15.1
Estimates for labour required for three cocoa management systems

Management system	Type of practice		Man–days /ha/year[1]
Extensive	irregular pod harvesting and breaking		15
	fermenting & drying		4
		Total	19
Minimal	regular pod harvesting and breaking		22
	fermenting & drying		5
	occasional manual weed control		3
	chupon removal		4
		Total	34
Intensive	regular pod harvesting and breaking		30
	fermenting and drying		6
	regular manual weed control		10
	structural pruning		8
	chupon removal		4
	insect pest control (chemical)		4
	phytosanitation of disease(s)[2]		–
		Total	62

[1] the man–day figures are estimates, based on Wood & Lass (1985) with reference to Latin America; [2] labour for phytosanitation is discussed below.

EXISTING PLANTATIONS

Cocoa is produced in a range of agricultural systems, that differ in the levels of inputs injected by the farmer. The simplest method consists of planting the crop, allowing the trees to develop naturally and collecting healthy pods as they ripen. Alternatively, substantial inputs can be applied to increase pod yields, and intervention can be as frequent and intensive as the resources and initiative of the farmer will allow. Three management systems have been chosen from the range of possible options available to a farmer, in order to illustrate the different scale of labour inputs required (Table 15.1). It is acknowledged that the world market price of cocoa at peak harvest–time has a major impact on the level of inputs applied by the farmer, and that a given producer will not always apply the same management system year after year.

It is extremely unlikely that farmers from "Extensive" or "Minimal" systems would willingly adopt any labour– or cost–intensive practices designed to increase production, as they do not invest in practices designed to achieve high yields. "Extensive" system farmers would also lose a certain number of pods through infrequent harvesting due to over–ripeness. "Minimal" system farmers are unlikely to have sufficiently high production to justify attempts to avoid disease losses. The following sections refer predominantly to "Intensive" systems, under which farmers are likely to invest resources on improvement or maintenance of productivity provided there is sufficient economic return.

General Crop Management

Techniques of crop management that will facilitate disease control need to be defined in conjunction with particular schedules for practices specifically aimed at reducing disease. Existing management schedules may need to be modified in order to improve the control of witches' broom.

Maintenance pruning

The importance of timely structural and maintenance pruning to achieve short easily-managed trees was stressed above in relation to new plantings. Maintenance pruning is equally important in mature plantations, and should be completed frequently to prevent basal chupons and canopy chupons from becoming too large. Chupons can also be removed during pod harvesting. Plantations which already exceed 5 m will be dealt with in the section of this chapter on rehabilitation.

Broom removal accompanied by structural pruning at some IWBP sites was found to cause an increase of about 20% in the total number of pods set independent of the proportion of pods lost to disease(s) when compared to unpruned plots. Light maintenance pruning is also known to have a stimulating effect on pod set in mature cocoa.

Harvest interval

Manual labour is often a finite resource that is fully occupied during the harvest period(s), because of the need for harvesting, breaking, fermentation and drying. Any practices aimed at the control/management of witches' broom will involve substantial labour, and should ideally be scheduled outside the harvest periods of cocoa or other crops of local importance. In sites with a clear dry season, the first and often larger harvest is collected in the late rainy season and early dry season. The second harvest, if it occurs, is composed of pods that set and develop during the dry season so that they mature in the early part of the rainy season.

Density of shade

Cocoa agronomists have long been concerned about the optimum level of overhead shade in a given situation, and opinions still differ. The effect of shade on cocoa disease adds another dimension but unfortunately there is little experimental information on which to base recommendations. Dense shade appears to reduce the number of basidiocarps produced on diseased tissues by prolonging waterlogging of these tissues, which allows saprophytic competitors of *Crinipellis perniciosa* to dominate. Shade also suppresses flushing and flowering, and fewer infection courts would give rise to fewer brooms per tree in comparison to an unshaded neighbouring planting similar in all other respects. The disadvantage of dense shade is that it may also reduce pod production and favour diseases such as *Phytophthora* pod rot which thrive in humid environments.

Irrigation

Trickle or flood irrigation is clearly preferable to spray irrigation, as the latter could induce basidiocarp production on diseased tissues. Irrigation has a stimulatory effect on flowering and pod production, and could be used to promote disease escape through out of season cropping. However, irrigation is not a regular feature of cocoa cultivation in the areas under study.

Practicality of phytosanitation

The most accessible point at which to interfere with the disease cycle remains the production of inoculum, and the key management method is the removal of potential sources of inoculum to sanitize the planting. The following sections refer only to managed plantations with canopies less than 3.5–4 m tall. "Extensive" cocoa over 5 m tall needs other treatments which are presented in the section on rehabilitation. In those latter cases, phytosanitation must be preceded by or combined with other practices.

Phytosanitation is generally practicable in well–managed plantations provided that recommended guidelines are followed thoughtfully. However, reductions in pod loss may not be significant if neighbouring areas of cocoa are not sanitized, and this is considered in the next section. There are some specific cases where phytosanitation guidelines need to be qualified.

The amount of pod loss to witches' broom in the majority of cocoa in Grenada and Trinidad does not justify action to control the disease. The underlying reason for low disease incidence is only clearly identifiable in Trinidad, where disease was reduced by the introduction of resistant varieties. The principal threat to cocoa production from witches' broom is indirectly through accumulated cushion infection and death. There is no direct evidence that reduced cushion activity decreases pod set, but the risk should be avoided where the potential for cushion disease incidence is high, and so phytosanitation is recommended even in areas of low incidence of cushion disease.

Feasibility of Phytosanitation

Scientific feasibility

The IWBP phytosanitation and gradient studies showed that careful broom removal reduces disease incidence dramatically in some, but not all, situations. Successful phytosanitation is jeopardized when inoculum from adjacent or nearby farms reaches the sanitized area in sufficient quantities to cause appreciable pod loss and abundant infections on shoots and flower cushions. The amount of incoming inoculum will depend on several factors, including the proximity and extent of neighbouring diseased cocoa, its amount of diseased tissue, and local air movements. It is not possible to predict precisely the outcome of sanitation for a particular situation, but some guidelines can be given. It is assumed that the removal of brooms is efficient and at the appropriate time for the cocoa in question, and that the pruned material is dealt with in a suitable way.

The distribution of cocoa varies greatly across regions and countries, from almost continuous to sporadic. In all plantings of cocoa where phytosanitation is practiced in the surrounding cocoa there is a high probability that pod losses can be reduced substantially. In discontinuous plantings (farms separated by at least several hundred metres) where phytosanitation is not practiced in the neighbouring cocoa the chances of achieving a successful reduction in pod losses are moderately strong. Prediction of any reductions in pod losses is more difficult for a sanitized field of cocoa surrounded by continuous plantations where brooms are not removed. The probability of significant reductions will be smaller in areas where disease incidence is high, but the possibilities for yield increases will be greater than in low incidence areas.

Financial feasibility

This subject was considered in a model for isolated cocoa developed by Rudgard & Andebrhan (1988), which was used to examine the likely thresholds of production required for adequate financial return from phytosanitation of a heavily infected plantation. The driving variables were considered to be the number of brooms per tree and production per hectare, and the factors under the farmers' control were the efficiency of the broom removal and the rate of removal per man-day. The model used data on the numbers of man-days of labour usage required to clean trees with different numbers of brooms, calculated under conditions in the Brazilian Amazon.

TABLE 15.2
Estimates of labour required for broom removal

Number of brooms mean/tree	Man-days/ha
5	13
10	17
20	26
50	35
100	50
200	63

The removal of about 30 brooms/tree would require half the annual labour requirement for intensive management of a healthy plantation (Table 15.2). The model shows the very high labour requirement for sanitation removal in plantings with large numbers of brooms. The average number of brooms should be kept low in intensively managed farms with efficient and sustained sanitation. Heavy witches' broom infestations would require a different approach which is considered in this chapter under the title of rehabilitation.

The independent variables in the Rudgard and Andebrhan model were the rate of pay per man-day (C) and the price to the farmer of dry cocoa per kg (p). The ratio C/p had a large influence over the profitability of the practice, both variables are out of the farmers' control. The ratio C/p for Rondonia, Brazil, increased from about 3.6 in 1985–6 to about 8.5 in 1991, thereby making it totally unprofitable to remove any brooms, even in plantations with yields of over 2 t/ha.

Techniques for Phytosanitation

Removal of brooms and diseased pods

The tool used to cut off brooms will depend on the type of implement available, the ease of access and height of the brooms from the ground. The most frequently used tool is a hand-sharpened blade on a pole (with or without an extension), but occasionally secateurs on poles or long-handled pruners are employed. The pole-mounted blade has the advantage of being cheap and durable, and it is already used by farmers for pod harvests. In cases where removal of individual brooms is feasible, vegetative brooms should be cut off at least 15–20 cm below the point of infection. Diseased material on cushions should be carefully removed by cutting it off as close as possible to the bark. Diseased pods together with their peduncles should be removed whenever healthy pods are harvested.

Efficiency of removal

Diseased tissues are distributed through the branches and canopy of the trees, with the bulk of brooms (both cushion and vegetative) and diseased pods generally being found in the high canopy, with only relatively few on the trunk and lower branches. Efficiency of broom removal, and the time taken, will depend on the ease of access to the diseased tissues, and this is related to the height of the trees (see above) and the architecture of the canopy. The removal must be as thorough as possible to have any chance of success.

Fate of removed brooms and diseased pods

It is important to ensure that all removed brooms and diseased pods reach the ground and do not remain suspended on branches within the canopy. Elimination of the diseased tissues from the plantation after pruning is optional in some circumstances, but necessary in others. The practice is recommended if annual sanitation has lapsed for more than one year, because older brooms on the ground could produce basidiocarps within a few days after the start of the rains if not removed. Where sanitation is regular, pruned tissues can be left in the plantation, provided that the following precautions are taken:

- Removed diseased material should not be heaped or left uncovered on the ground in plantations;
- Leaf-litter should be allowed to accumulate naturally, or placed over the removed diseased material.

It is recommended that removed brooms and diseased pods should be eliminated from the plantation in areas where basidiocarps are produced on first-year brooms on the ground. Broom removals can be timed to take advantage of heavy leaf fall before and during shoot flushing periods, so that brooms on the ground are covered as quickly as possible.

Diseased pods need to be separated from healthy pods before the latter are processed in order not to affect the quality of the beans produced.

If phytosanitary removal is combined with structural pruning, it is likely that branches with brooms will be cut off the tree, in which case such material should be cut into small pieces. The material can then be spread on the ground so that it is all in contact with the leaf-litter as this speeds up decomposition.

Other pod diseases

If witches' broom is controlled, then the levels of losses to other pod diseases in the plantation might increase and thereby reduce the economic benefit of successful witches' broom control. This could happen with *Moniliophthora roreri* in Ecuador or Colombia, and in such cases measures for management of the second disease are also necessary if yield increases are to be secured.

Timing of Phytosanitation

Phytosanitation must be correctly timed, and the choice of times as outlined at the end of each national epidemiology study (Chapters 5–10) was based on the durations and the timings of the climatic and disease cycles. In many cases, it was proposed that additional (secondary) removals would be required to supplement a main (primary) sanitation pruning to ensure effective disease management.

Phytosanitation should not be attempted where it is not possible to complete the primary removal at the recommended time. No secondary removal should be applied without a primary removal. Sanitation is most effective where it is routinely practised every year, so that numbers of second-year brooms are kept to a minimum.

General principles

- Primary removal: a number of criteria for selection of the time of primary broom removal can be specified for all areas. Phenological data show that the periods of greatest incidence of brooms and the periods of minimum production of basidiocarps, coincide during the drier season. At this time, most of the diseased tissues are necrotic and easier to distinguish. The objective of the primary removal is to remove diseased material before the onset of the rains (and thus the onset of basidiocarp production), particularly as the limited amount of sporophore production in the early part of the rainy season can be significant. In areas with no distinguishable drier season, the primary removal is best timed 1–2 months after the end of the main period of broom formation.

A delayed dry season or early onset of rains requires respectively a late or an early start in the application of the schedules. Times are presented for each of the areas involved in IWBP as guidelines only.

- Secondary removal: a secondary removal is required in some areas to cope with the late-emerging brooms and those missed in the primary removal, and in most cases, to further reduce the number of productive brooms before the main period of pod set and development.

Regional recommendations

Within the general principles, it is possible to make recommendations for the individual IWBP sites, as summarized in Table 15.3. These recommendations are also made in the light of comments on the feasibility of sanitation.

Brazil – Amazon region. The dry season occurs at the same time throughout the Amazon Basin of Brazil, and the recommendations for the removal times are the same. The primary removal should be in the months of August/September, followed by the secondary removal 2–3 months later (in November/December).

TABLE 15.3
Regional guidelines for the timing of primary (P) and secondary (S) removals of diseased witches' broom materials from cocoa

	Months											
Country	J	F	M	A	M	J	J	A	S	O	N	D
Brazil[1]								P	P		S	S
Ecuador[2]							P	P	P		S	
Colombia		P	P				S					
Venezuela[3]			P									
Trinidad			P						S			
Grenada			P									

[1] Amazon region; [2] Coastal region; [3] Táchira State.

Colombia (Caldas and Llanos)/Venezuela (Táchira). In well-managed plantations where witches' broom is already present, the most appropriate time for the primary removal is the middle to end of the principal dry periods in February/March, and March/April for Táchira. A secondary removal is not recommended in Caldas and Táchira, but is recommended for June/July in the Llanos.

Grenada and Trinidad. Phytosanitation should particularly be applied in areas of high disease incidence, with a primary broom removal in March/April before the onset of the rainy season, with an optional secondary removal in September/October. The central region of Venezuela (Barlovento) should also follow this recommendation.

Ecuador – central coastal area. The primary removal can be combined with light structural pruning in July/August with the aim of stimulating flowering and pod development at that time to allow most pods to escape infection. In cases where no structural pruning is applied, the primary sanitation should be completed by the end of September. A secondary removal is required in November. Diseased pods should be removed every 7 days in order to reduce the incidence of moniliasis.

REHABILITATION

Rehabilitation of a plantation involves the application of treatments to increase the cocoa production of low-yielding trees. Low production can be due to age, mismanagement or lack of management, and can also be related to the cumulative effect of a chronic witches' broom attack that has never been treated. A large commitment of resources is required to rehabilitate any cocoa plantation, and the farmer is exposed to potentially heavy financial losses if treatments are applied incorrectly. Where witches' broom is present, application in single plantations may not be successful due to the risks from background disease pressure outlined above. Rehabilitation of a witches' broom infected plantation also needs to be associated with a commitment to regular maintenance and regular sanitation pruning to maintain any improvement in yield.

Treatments to rehabilitate plantings must be carefully planned in areas where witches' broom incidence is high, as the risks of failure are greater. High disease incidence can itself force farmers to change from intensive management, because cumulative cycles of broom formation cause heavier pod losses and reduce the farmer's income. Such difficulties are exacerbated in times of low cocoa prices.

Types of Treatment

The practices presented below are recommended from previous work on the disease, and from experience in the IWBP of existing cocoa types. Cocoa plantations are considered in one of two categories for the recommendation of treatments.

The first category has cocoa with a branch structure originating below 2 m. Such trees should receive a structural pruning to reduce height and improve access to canopy, followed by a sanitation pruning. Severely infected trees might have branches cut back drastically. The second category consists of cocoa that lacks any branch structure below 3–4 m and trees should be pollarded at between 0.5 m and 1.5 m from the ground to stimulate chupons for the regeneration of the tree. Bud grafting onto basal chupons is an additional possibility.

Plantations of cocoa with the seedling habit may contain individuals that are exceptionally heavily attacked by witches' broom, which may also be unproductive. Such trees could be eliminated and replaced by better material. In cases where the potential production of the existing cocoa is generally low, interplanting with improved seedlings followed by gradual removal of the old trees ("Turrialba Method") might be preferable to rehabilitation. Frequent protective fungicide sprays would be necessary for the seedlings in areas of high disease incidence until the trees become well established (Maddison, 1991).

Timing of Rehabilitation

The success, or failure, of these treatments is influenced by the time of year in which they are carried out. The most important consideration is the phenology of the recuperating trees and the microclimate to which the regenerating trees will be exposed. In shaded plantings where variations of microclimate are less violent, the selected pruning treatment should be applied towards the end of the rainy season. This allows the canopy to start reforming during the dry season and continue into the early rainy season when there is no risk of infection. The recommendation only needs to be adapted for unshaded plantings where daytime temperatures are excessive (over 32°C) in the dry season (Brazilian Amazon). In such cases, the selected pruning treatment should be applied at the middle/end of the dry season.

Fate of Removed Material

All the above treatments generate considerable quantities of infected material pruned onto the ground within the plantation, and much is capable of producing basidiocarps. All the branches, foliage and brooms should be cut up, making a compact layer of organic matter that will be quickly covered with fallen leaves and decompose rapidly. Mineral oil (at 40–60 l/ha) can then be sprayed onto the cut material on the ground to inhibit sporulation on the brooms. This is not an expensive chemical, and is widely available.

Effects on Pod Production

These treatments are intended to increase pod production in the medium term, but will immediately reduce pod production as they involve the removal of either part, or most, of the fruit-bearing wood and/or the leaf area of the canopy. The beneficial effects of such practices on yield and the time taken for recovery are being investigated in some of the countries collaborating in IWBP. Early results are encouraging, and indicate that about 1 to 3 years are required for the two recommended rehabilitation pruning methods. The "Turrialba Method" involves no appreciable yield loss if carefully managed.

CONCLUSIONS

The IWBP has shown that phytosanitation has its place in the management of cocoa infected with witches' broom in certain situations. The project has provided evidence that the practice (correctly applied in intensively managed plantations) will reduce disease incidence on all parts of the cocoa tree in most areas within the present distribution of the disease. The economic margins vary according to disease pressure and the completeness of the removal.

The world market price of cocoa has a significant influence on the return due to the application of management practices, as farmers will gradually abandon more demanding activities as margins tighten. The effect on yield of alternate periods with and without phytosanitation still needs to be studied, but it would appear that provided pruning to maintain architecture is continued and that trees still have adequate numbers of healthy flower cushions, it is possible to bring farms back into reasonable bearing through phytosanitation.

There is no merit in recommending phytosanitation in the large areas of extensively managed cocoa, as it is a labour intensive practice, and therefore unlikely to be adopted by low–input farmers. The answer lies in a longer–term approach, unfavourable to most managers of research. Germplasm management by selection and breeding holds the key to future treatment of witches' broom disease. Durable resistance is the only sustainable method of increasing production for the section of the cocoa–growing community that will continue to farm the crop with few resources. Resistance will be discussed more fully in the next chapter.

Chapter 16

FUTURE PROSPECTS FOR IMPROVEMENTS IN DISEASE MANAGEMENT

S.A. Rudgard, T. Andebrhan, A.C. Maddison & R.A. Schmidt

INTRODUCTION

The previous chapter dealt with recommendations for the management of witches' broom. However, these methods are not universally viable and are not sustainable in times of adverse economic conditions even for those farmers who are willing to manage their crop intensively. It was also stated in the conclusion of Chapter 15 that the large number of low-input farmers would not consider the application of any treatments for the management of witches' broom. Long-term solutions need to be found for economically stable maintenance of tolerably healthy productive cocoa, and particularly for areas of high witches' broom incidence where sanitation is at best only marginally applicable. The IWBP did not have the resources to consider such problems, but thought was given to the strategies required to research and develop alternative techniques.

Two alternative management methods apart from sanitation are chemicals and disease resistance. There has been a significant amount of research on both these topics, but much development is required before they can be applied. All three methods could also be mutually beneficial. This chapter covers some areas where research is required to improve phytosanitation, briefly considers the implications of biological control, and outlines strategies for development of chemical control and resistance, the latter formulated by an IWBP Working Group. The scientists involved in the project all considered that resistance offers the most economically sustainable management strategy, and the greatest emphasis is given to this topic.

PHYTOSANITATION

Phytosanitation may be an essential practice to augment other strategies. However, as previously stated, it would be possible to greatly facilitate the removal of brooms by the development of lower-growing trees, or cocoa varieties that have a shorter flowering period and which do not have the propensity to form vegetative brooms on cushions. Furthermore breeders/geneticists could also contribute to reducing the costs of broom removal through

Disease Management in Cocoa: Comparative epidemiology of witches' broom. Edited by S.A. Rudgard, A.C. Maddison and T. Andebrhan. Published in 1993 by Chapman & Hall, London. ISBN 0 412 58190 6

development of resistant varieties that form fewer brooms.

The previous chapter considered the application of phytosanitation within the limitations of existing tools, but the IWBP also spent some time investigating the possibilities of improving the implements used to remove brooms. Mechanical pruning devices in use in orchard and plantation crops were considered, and in particular the pneumatic shears widely used in temperate orchard crops. At the moment, the systems available are relatively expensive and their size and lack of mobility would be unsuited to cocoa farming systems that are often without easy access. It would be difficult to replace a simple knife blade with such a comparativly high-technology tool, unless it was locally made and powered by a portable compressor.

There are aspects of phytosanitation which merit further research or were not considered during the IWBP. Greater knowledge of disease spread in a wider range of environments and over more seasons would be advantageous, particularly with reference to the importance of background inoculum to plantations where brooms have been removed. Further studies on inter-plot interference and experimental design would have to be completed, if any comparison of timing and frequency of broom removal was being contemplated within one plantation.

BIOLOGICAL CONTROL

The review of the pathosystem in Chapter 2 showed that relatively little research had been completed on this subject in relation to the other areas of disease management, and all of it had been published early in the life of the IWBP. Biological control was not identified as a separate strategy in Chapter 2, as its development in relation to witches' broom had generally been together with sanitation or chemical control. Classical biological control by the introduction/application of a living natural enemy of *Crinipellis perniciosa* has been studied and antagonists in the form of hyperparasitic fungi have been identified. They either attack basidiocarps or colonize dead diseased tissue in competition with *C. perniciosa*. However, no organism has been found with sufficiently aggressive antagonistic activity under field conditions to merit further development. However, the management of removed material recommended in Chapter 15 aims to exploit a type of biological control, through modification of the microclimate in the litter layer to favour microbiological decomposition of brooms.

Culture extracts of certain antagonistic fungi have been shown to have inhibitory activity against basidiospore germination and saprophytic growth of *C. perniciosa*, but such extracts seem to be merely fungistatic active ingredients that would have to be bulk manufactured and formulated for widescale use. Similar difficulties in reaching brooms with sprays of such extracts would exist to those of reaching canopy pods with fungicide sprays.

FUNGICIDES

Feasibility

Several reviews of the use of chemicals for the control witches' broom disease, were discussed in Chapter 2, but the general picture is one of theoretical promise which does not bring success in practice. For instance, it has been shown by several workers (Laker & Rudgard, 1989) that fungicides can significantly reduce pod disease incidence, although other workers have found no effect using the same main ingredient. The dose, interval and number of sprays have all been found to affect activity. Commercial application of fungicides for the control of witches' broom has not been adopted in any cocoa-producing country, because increases in production have not given sufficient economic return to motivate farmers into applying the treatments. The two main strategies identified by Laker & Rudgard (1989) as holding realistic promise for development are:

- Protection of pods with a broad-spectrum residual fungicide such as copper;
- Inhibition of broom formation and eradication of mycelium in shoots and flowers using a systemic fungicide.

The factors that govern the economic viability of the use of fungicides are similar to those controlling the return on phytosanitation. The price of cocoa to the farmer is still the critical variable, and, of course, higher inputs are necessary for the use of chemicals because applications have to be frequent and extra costs are incurred in labour, equipment and fungicide. Production would need to be increased in order to justify fungicide application, as the proportion of pods that are healthy and/or an overall increase in the number of pods produced. It is likely that chemicals would only be used in plantations with reasonable overall production and appreciable pod losses to witches' broom. The levels of farmers' inputs in the management systems discussed in Chapter 15 are still applicable here, and it is likely that fungicides would only be applied where management was intensive.

Protective Sprays

Spray equipment and residue retention

Economic protection of developing pods could be achieved by sprays in the main inoculum period, if their residual activity could be improved. However, cocoa has a very different spatial arrangement of spray targets to the majority of crops, and spray equipment typically used in cocoa has been developed for general purposes and is actually not well suited to cocoa. This has been a significant inhibitory factor to the development of fungicide usage in this crop (for both witches' broom and other pathogens). Pods are relatively small targets which are dispersed throughout the lower architecture and canopy, up to a height of several metres. Conventional hand-operated knapsack sprayers do not have the ability to reach the heights required, and motorized mist-blowers produce a cone of spray and much of the chemical misses the target as well as being unevenly distributed over the surface of the pods. The greater part of chemicals sprayed in cocoa plantations lands on the litter layer. Chemical is also lost when pod surfaces are saturated with spray and run-off occurs.

Residues are eroded by rain and canopy drip, although some redistribution by splash can occur between pods in the canopy. Pods are most susceptible in the period of greatest growth, which means that fungicide residues are diluted by surface expansion. Both the above factors lead to loss or dilution of active ingredient and result in the breakdown of disease control. Alternative methods of treatment have been found where, with only one application, surface concentrations of protectant fungicide were maintained above lethal doses for sufficient periods (Pereira, 1985; Sreenivasan *et al.*, 1990), and it is possible that such techniques will be made commercially viable.

Timing of sprays

Correct timing is essential if treatments are to be most effective. Epidemiological data can provide information on the period of highest risk to pods in terms of the presence of basidiocarps, and also the periods with the greatest numbers of susceptible pods. The comparative epidemiology experiment enables researchers to identify or confirm the most effective times to apply protective chemicals for control of pod disease from witches' broom.

Systemic/Eradicant Compounds

Systemic compounds for protection of vegetative or flower cushion activity might be sprayed on to the canopy, and eradicant compounds could be used to kill developing mycelium in green brooms. Knowledge of the activity of each infection court would be fundamental to the correct use of fungicides. Compounds have been found that translocate sufficiently well with cocoa plants to allow their use in soil applications in young trees. Such compounds might cause problems of residues in the cocoa beans. If those reservations were dealt with satisfactorily, applications would have to be timed slightly earlier than foliar sprays, to allow time for the chemical to be translocated to growing apices.

Compounds have also been proposed for use as antisporulants to be sprayed on to brooms, whether on the tree or the ground (e.g. mineral oil, Chapter 15). Research has identified the danger that certain active ingredients can stimulate greater basidiocarp production, and Laker & Rudgard (1989) did not attribute a high priority to the devlopment of antisporulants.

DISEASE RESISTANCE AND BREEDING

While the IWBP has provided recommendations for the management of witches' broom disease by phytosanitation, it was recognized that improvements in long–term control of witches' broom can best be achieved by the use of resistant cocoa planting material and the development of such material is urgently required. The rest of this chapter consists of a text developed by a Working Group of the IWBP, outlining a research plan that would be likely to lead to the development of cocoa materials with resistance to the disease.

Background

The search for resistance to witches' broom began in earnest in the late 1930s when F.J. Pound collected apparently immune cocoa from commercial estates in Ecuador (called "refractario" cocoa) and wild cocoa from the forest in Peru (the Upper Amazon material). This germplasm was established in Trinidad where the incidence of witches' broom was high and susceptibility/resistance of individual trees could be assessed by the level of broom formation. The emphasis then was to seek immune individuals and this led to the selection of two clones, Sca 6 and Sca 12, from which crosses were made subsequently. While such hybrids had some measure of success in Trinidad, they became severely infected in Ecuador where it is now known that a different pathogenic form of *C. perniciosa* exists. Success in breeding for resistance to witches' broom has thus been limited and remains so.

However, the most effective measure for disease control in Trinidad has been the breeding of the "Trinidad Selected Hybrid" (TSH) family, and adoption of the TSH clone set in commercial farms over the traditional ICS materials which were planted up to 1960. The resistance of the TSH materials has never been established in laboratory tests, but has been shown as a field response. The early insistence on immunity has probably caused other types of resistance (horizontal resistance or HR) to be ignored, or worse, discarded, and the absence of a valid screening test for large populations has restricted selection of parental material and subsequent progeny.

Pound's expeditions showed that the Amazon basin was a vast reservoir of *Theobroma cacao* germplasm and other collections have been made over the past 50 years in Brazil, Colombia, Ecuador, Peru and Venezuela. The available genetic resources at an international level are considerable and most notably are present in germplasm collections in Brazil, Ecuador and Trinidad with less diverse collections elsewhere. This outline proposal would include the use of a wide genepool in a search for durable resistance to witches' broom.

Disease Resistance

Some resistance to *C. perniciosa* exists in cultivated cocoa but little is known about its mechanism. Danquah (1986) showed that pathotypes from both Trinidad and Pichilingue (Ecuador) could infect seedlings with Sca 6 as one parent but the Trinidad isolate grew less well, induced less swelling and caused host cells to die more rapidly. In cultivated cocoa, there appears to be a complete range in severity of symptoms depending on pathotype and cocoa clone. The range of suceptibility evident in the Trinidad population at the time led Baker & Crowdy (1944) to comment, "[this]... would be expected if resistance was due to a large number of genes of small individual effect occurring at random in the unselected population. ... In South America natural selection for resistance has operated ... for a considerable period. For this reason ... it is to be expected that resistant trees would be more common or more highly resistant in the Amazon Valley than elsewhere". This has proved to be so. Some collectors have found trees without witches' broom, but whether this is due to immunity conferred genetically or disease escape through lack of contact between inoculum and susceptible tissue is not known. An evaluation of 281 accessions (LCT–EEN series), collected from the Upper Amazon and exposed to natural inoculum at San Carlos, Ecuador, favours the latter hypothesis (escape). Preliminary results indicated a range of susceptibility to witches' broom with the frequency of infected pods amongst the accessions apparently approximating to a normal distribution (Allen & Cabrera, 1988). Some accessions with relatively little witches' broom and high yields are of obvious interest but, equally, so are those which yield well despite high rates of infection.

The picture which emerges from considering pathosystems generally is that vertical resistance (VR) is unlikely in tropical perennials such as cocoa (Robinson, 1976). Cocoa pathosystems appear continuous, and no gene–for–gene relationships are known (Robinson, 1987). That this applies to the *C. perniciosa/T. cacao* pathosystem is supported by the screening of cultivated cocoa in Trinidad and the evaluation of wild cocoa in the LCT–EEN series in Ecuador. The concept that the *C. perniciosa/T. cacao* pathosystem is essentially a horizontal pathosystem does not exclude the possibility of differential interactions like that demonstrated for the Trinidad and Pichilingue isolates with cocoa of Sca 6 parentage (Wheeler & Mepsted, 1988). There is some evidence that the apparent immunity of Sca 6 can be genetically diluted by hybridization with other cocoa genotypes (Thorold, 1975). Such considerations led to a fundamental change in ideas about breeding for resistance to witches' broom (Bartley, 1977; presentation by A. Kennedy in Lass & Rudgard, 1987). If HR exists, as the above features suggest, then a breeding strategy could be devised to accumulate and incorporate HR genes into cocoa varieties to avoid the potential danger of a "breakdown" of resistance sometimes associated with VR in a long-lived perennial tree crop plant. This would increase the possibility of obtaining long lasting resistance, so essential for a perennial crop. It would be important to this concept of durable resistance that in the screening process, a large and diverse base of germplasm be examined initially, that VR, if it were to exist, be eliminated, and that an aggressive pathotype be used. Testing should be accomplished at several geographically and thus environmentally distinct locations, and exchange of germplasm would be essential in the early stages of testing. This strategy was developed by an IWBP Working Group, and is reproduced here.

Resistance Breeding Objectives

The four primary objectives of a witches' broom resistance breeding programme would be:

- To select base parental populations with horizontal resistance to witches' broom from germplasm collections in Brazil, Ecuador and Trinidad;
- To exchange these parental populations so that all possible genes are made available at each site;
- To develop new mass screening techniques involving the artificial inoculation of young seedlings, thus reducing the time taken to select potentially useful material;
- To utilize the screening techniques in a new breeding strategy to select populations with improved horizontal resistance to witches' broom.

Breeding Strategy

The breeding strategy would not necessarily intend to develop new commercial varieties in its early phases. Once a breeding population carries a high probability that its progeny are horizontally resistant to the disease, different gene pools could be introduced to include other important characters. The breeding strategy would be applicable in a general sense to any characters under selection, such that parallel populations could be developed for locally significant characteristics with the intention of combining these populations at a later stage of variety development.

Development of a high level of horizontal resistance needs to be based on selection of a starting population that has a wide genetic base. This base parental population (P_0) would form the starting point for a recurrent mass selection programme. The cycle period for such a scheme would be 5 years, each cycle comprising parental selection, mating, greenhouse screening of progeny as seedlings, planting survivors in the field for further testing and selection of the next parental generation (Table 16.1).

The mating programme should be such that at each cycle a genetic analysis could be made for the inheritance of resistance (or for any other character of interest). There would also be a measurable response to selection at each cycle of the programme.

The starting point for the programme would be essentially the same at each proposed site. However, the primary germplasm is very different at each site comprising collections from different geographical areas of South America. The collection in Brazil is essentially from the lower Amazon basin, in Ecuador from the headwaters of the Amazon (the centre of diversity of *T. cacao*) and in Trinidad the collection is mainly wild cocoa from Ecuador and Peru or selections from old commercial plantings in Ecuador and Trinidad. Together these three collections hold a sample germplasm that spans the geographical range of *T. cacao* with the exception of the Central American "criollo" population.

TABLE 16.1
Representation of a breeding cycle for development of resistance to witches' broom disease of cocoa

Stage	Activity
1.	Select 50 base parents (P_0) from each of the three germplasm collections.
2.	Exchange these with 50 from other two germplasm collections.
3.	Test the 100 imported parents by artificial inoculation with local isolates of *C. perniciosa*.
4.	Plant survivors in field for flowering and inclusion in later crossing.
5.	Divide P_0 into 25 seed parents (SP) and 20 pollen parents (PP) at random.
6.	Cross in NCM 11 design[1] to give 500 crosses.
7.	Collect 30 seeds per cross and mark each batch of 30 with its SP and PP origin.
8.	Exchange spare seed[2] between sites for screening.
9.	Mass screen of 15,000 seedlings from above.
10.	Plant survivors an area of high disease pressure.
11.	Observe reaction to witches' broom for 4 years.
12.	Make final selection[3] of P_1 for parents for next generation.
13.	Repeat from 2, include exchange parents and selections from exchange seed.

[1] NCM11 design is where each SP is mated with each of the 20 PP. This allows for genetic analysis and precise quantification of progress; [2] exchange of seed allows for estimates of site (pathotype) interaction; [3] exact number of parents may vary from site to site.

The efficiency of the programme would be greatly enhanced by exchange of each parental level population, because scientists at the three proposed sites should work with at least three different pathogenic variants of *C. perniciosa* and very different germplasm populations. In this way usable genes in Trinidad, for example, undetected by the local variant might be made apparent at other testing sites. Free exchange, therefore, would make maximal use of all available germplasm and to some extent guard against selection for resistance to local variants with its inherent danger of developing vertical types of resistance.

Screening for Resistance

A successful testing programme depends on an appropriate inoculation technique. This technique would need to be consistent, provide for control of inoculum concentration, and be easily and rapidly applied to large quantities of cocoa material. The method should mimic natural (field) inoculation as closely as possible. Timely production and availability of inoculum would also be important. Additionally, control of plant growth pre- and post-inoculation and of the ambient environmental conditions would be necessary[1].

Selection of parental populations would be based initially on available data on vegetative, cushion and fruit infection in germplasm collections. Screening of seedling populations would be based on vegetative infections for speed and convenience, and also canopy infection is the principal generator of witches' broom epidemics. Field tests would indicate, amongst other things, the levels of resistance to cushion and fruit infection.

Benefits

Among the benefits of this programme to all cocoa producers may be listed:

- If the pathogen ever spread to new areas where it is presently absent, then the cocoa populations arising from this programme would be immediately available for planting in these newly infested areas. This would be a most necessary insurance policy for those producers not presently affected by the disease;
- Once the technique and system of breeding for resistance to witches' broom has been developed and proven then it may be applied to incorporate resistance to other diseases where horizontal resistance has been demonstrated or to incorporate other agronomic or economic characters. For example, it seems likely that horizontal resistance for vascular streak dieback will be proven in due course;
- It is believed that this programme could produce a new generation of high yielding hybrids which could benefit the whole cocoa industry by having the possibility of much greater productivity in terms of cocoa grown per unit land area. If this project were funded internationally then these hybrids would clearly be available to the whole of the world cocoa community.

Among the benefits of this programme to cocoa growing areas presently affected by the disease may be listed:

- Cocoa populations arising from this programme would have improved horizontal resistance to witches' broom disease and would thus have greatly improved yield prospects when planted in areas where the disease is presently causing such devastation;
- The incorporation of horizontal resistance to the pathogen into a range of planting material would be complementary to the other approaches to disease management developed in IWBP, such that a genuine system of integrated disease management may be adopted. This should be encouraged since it would be both environmentally and economically desirable as it would minimize the use of pesticides.

In view of the importance of this programme, support by both cocoa producers and cocoa consumers is recommended. The opportunities exist for substantial improvements in productivity (that is to say higher production per unit area of land allocated to the crop). This would be to the benefit of the world cocoa industry.

---

[1]Editors' Note: Since the Working Group produced this document, a method for mass screening has been developed by Purdy and Schmidt in Gainesville, and installed in two locations.

Appendix I

IWBP COMMITTEES

PROJECT MANAGEMENT COMMITTEE (PMC)

The terms of reference of the PMC were established in July 1985 and were as follows:

This Committee:

- Approved scientific objectives of IWBP;
- Approved recruitment of IWBP staff;
- Received contributions from participating member Associations;
- Approved budgets;
- Received 6 monthly statements of funds from IOCCC;
- Monitored progress towards scientific objectives;
- Reported annually to all participating Associations;
- Received and approved reports by Executive Chairman.

The members were nominees of the Member Associations of IOCCC who contributed funds to IWBP.

P.H. Leurquin, Association Belge Des Chocolatiers, Avenue de Cortenbergh 172, B–1040 Brussels, Belgium

B.K. Matlick, Hershey Foods Corporation, 19 East Chocolate Avenue, Hershey, PA 17033, USA

E.T. Beauchamp, BCCCA, 11 Green Street, London, W1Y 3RF, UK

J.N. Wintgens, Agricultural Services, Nestec, Avenue Nestle 55, CH–1800 Vevey, Switzerland

J. Colaneri, Chambre Syndicale Nationale Des Chocolatiers, Rue de Rivoli 194, F 75001, Paris, France

K. Schutze, Bundersverband der Deutschen, Sussmannstrasse 4–6, D–53 Bonn, Germany

A. Arnoudse, Cacao Fabriek de Zaan bv, P O Box 2, NL 1540, AA Koog A/D Zaan, Holland

A.P. Williamson, Cadbury Ltd, P O Box 12, Bournville, Birmingham, B30 2LU, UK

D. Kuster, Chocosuisse UFSC, Munzgraben 6, 3000 Berne 7, Switzerland

R. Bralsford, Cadbury Ltd, P O Box 12, Bournville, Birmingham, B30 2LU, UK

D. Duevel, Institut für Angewandte Botanik der Universitaet Hamburg, Willistrasse 20, D–2000 Hamburg 60, Germany

Th. Van der Warden, Nederlandse Cacao en Cacao Produkten Vereniging, Nigenburg 75, NL–1080 GE, Amsterdam Buitenveldert, The Netherlands

G. Holmqvist, Svenska Choklad–Konfektry–Och, Kexfabrikanter, Storgatan 19, Box 5501, S–11485 Stockholm, Sweden
H. Flotzinger, Mars Europe Ltd, 30 Dundee Road, Slough, Berkshire, SL1 4LG, UK
P.O. Werling, AB Marabou, S–172 85 Sundbyberg, Sweden
J. Van Kan, IOCCC, Avenue de Cortenbergh 172, Brussels, Belgium

PROJECT SCIENTIFIC COMMITTEE (PSC)

The terms of reference for the PSC were established in July 1985 and were as follows:

This Committee:

- Approved and agreed scientific objectives;
- Approved appointments of staff;
- Approved detailed work programme;
- Annually reviewed progress against objectives and against work programmes and recommended changes as necessary;
- Approved selection of Project Liaison Officer.

The members of this Committee are listed below, with an indication of when they were members in parentheses.

P.T. Alvim, Advisor for Scientific and International Affairs, CEPLAC, Caixa Postal 7, Itabuna, 45600, Bahia, Brazil (from May 1987)
T. Andebrhan, IWBP Coordinator, CEPLAC/DEPEA, Caixa Postal 1801, CEP: 66.000, Belem, Pará, Brazil (from May 1987)
F. Aranzazu H., IWBP Coordinator, ICA, Apartado Aereo 876, Manizales, Caldas, Colombia (from May 1987)
P. Buriticá, Subgerente Investigaci6n, Instituto Colombiano Agropecuário, ICA, Bogotá, Colombia (from July 1985 to September 1987)
R.A. Lass, Executive Chairman IWBP, Cadbury Ltd, Bournville, Birmingham, B30 2LU, UK (throughout)
A.C. Maddison, IWBP Coordinator, 20a Moor Lane, Chessington, Surrey, KT9 1BW, UK (from May 1987)
L.H. Purdy, Chairman PSC, IFAS, Plant Pathology Department, University of Florida, Gainesville, Florida 32611, USA (throughout)
H. Reyes E, CENIAP, IWBP Coordinator, Apartado Postal 4653, Zona Postal 2101, Maracay, Estado Aragua, Venezuela (from September 1987)
S.A. Rudgard, Project Liaison Officer, CAB International, Development Services, Wallingford, Oxon, OX10 8DE, UK (throughout)
R.A. Schmidt, Dept of Forestry, School of Forest Resources and Conservation IFAS, University of Florida, Gainesville, Florida 32611, USA (throughout)
T.N. Sreenivasan, IWBP Coordinator, Cocoa Research Unit, University of the West Indies, St Augustine, Trinidad, West Indies (throughout)
C. Suárez C., INIAP, Plant Pathologist, Pichilingue, Apartado 24, Quevedo, Ecuador (throughout)

H. Toxopeus, Nassauweg 14, 6703 CH Wageningen, The Netherlands (from July 1985 to May 1987)

B.E.J. Wheeler, Reader in Plant Pathology, Imperial College, Silwood Park, Ascot, Berks, SL5 7PY, UK (throughout)

DATA MANAGEMENT AND MONOGRAPH COMMITTEE (DMMC)

The terms of reference for the DMMC were established in November 1988 and were as follows:

Data Management

- To advise on the summaries, analyses, interpretation and presentation of the data from the Comparative Epidemiology Experiment;
- To review the summaries, analyses and interpretation of the above data;
- To present the summaries, analyses and interpretation of the above data at the 6th IWBP Workshop to be held in Florida;
- To prepare the chapter on the Comparative Epidemiology Experiment for the IWBP Monograph in a timely way;
- To prepare a draft discussion document on witches' broom disease management for PSC and for discussion at the 6th IWBP Workshop to be held in Florida and for finalisation at 7th Workshop;
- To catalogue the data from the Comparative Epidemiology Experiment for future reference and potential use by other scientists;
- To consider other project data as agreed.

Monograph

- To prepare an outline of the IWBP Monograph in January 1989 and to propose authors for chapters of the Monograph;
- To obtain and edit manuscripts for individual chapters of the Monograph;
- To advise on and facilitate the publication of the Monograph;
- To prepare a budget for ther preparation of the Monograph and present it to the Executive Chairman of IWBP.

The members of the DMMC were as follows (addresses are stated above):

T. Andebrhan, IWBP Coordinator
A.C. Maddison, IWBP Coordinator
S.A. Rudgard, Project Liaison Officer
R.A. Schmidt, Chairman

(The membership of DMMC did not change throughout the existence of the Committee.)

Appendix II

IWBP WORKSHOPS

A total of seven Workshops[1] were held, usually at research centres where work on the disease was being undertaken at that time. The Workshop details were as follows.

In early July 1985 scientific participants in IWBP met at Pichilingue, Ecuador, for the 1st IWBP Workshop to formulate the direction of the research programme on the epidemiology of *Crinipellis perniciosa* and agree experimental protocol for the comparative decisions on this were published in October 1985 in the "Report of the 1st IWBP Workshop". The collection of data in these epidemiology, chemical control and phytosanitation experiments was initiated at sites in six countries.

It was decided that it was necessary to consider the evaluation of candidate chemicals for disease control and a preliminary discussion was held at the 2nd IWBP Workshop held in Wageningen, The Netherlands in October 1985. This discussion was published in the "Report of the 2nd IWBP Workshop". Statements on the economic losses caused by the disease and on possible strategies for its control with chemicals were prepared for the meeting and are included in the report.

The 3rd IWBP Workshop was held from 7–12th July 1987 at Granja Luker, owned by Chocolateria Luker SA and situated near Manizales in Colombia. The experimental Protocol for the comparative epidemiology experiment (agreed in July 1985) was carefully and extensively reviewed at this Workshop and amended as necessary. An experimental Protocol for recording climatic data at sites where witches' broom disease was absent was also developed. Guidelines were drawn up for the evaluation of chemicals for control of witches' broom following the discussions at the 2nd Workshop.

At the 4th IWBP Workshop held in Belém, Brazil from 11–17th September 1987, the objective and philosophy of phytosanitation studies within IWBP was discussed and the experimental Protocol for the comparative phytosanitation experiment was established. Guidelines were also drawn up for conduct of Inoculum and/or Gradient Experiments.

---

[1] Copies of Workshop Reports are available free of charge from Dr J. Van Kan, IOCCC, Av. de Cortenbergh 172, Brussels 1040, Belgium

The 5th IWBP Workshop was held at the Asa Wright Nature Centre, Arima, Trinidad from 3rd–10th November 1988. The results at that time available from the epidemiology, inoculum and gradient experiments were carefully reviewed. Suggested methods of data management and presentation were discussed and some guidelines were produced. The experimental protocol for the comparative phytosanitation experiment was reviewed and some minor amendments were made.

The 6th IWBP Workshop was held at the University of Florida, Gainesville, USA, in April 1990. Discussions centred on the analysis of data from the comparative epidemiology experiment which had been so far completed. Preliminary ideas about control recommendations at some of the IWBP experimental sites were also elaborated in the light of those data together with the first results of the Comparative Phytosanitation Experiment.

The 7th (and final) IWBP Workshop was held at the University of Warwick in England in April 1991. This occasion was used to produce final recommendations for disease control, prepare the content of media material to present to farmers, and establish the direction of further scientific work on witches' broom disease of cocoa after the close of IWBP (assuming funds could be made available).

Appendix III

THESES PUBLISHED BY IWBP STUDENTS

Brazil[2]

Costa, J.C.B. (1992). Progresso da vassoura-de-bruxa em órgãos vegetativos do cacaueiro em Altanixa e Tomé-Açu. M.S. Thesis, Federal University of Viçosa.

Colombia

Cárdenas G., N. & Cárdenas S., A. (1991). Control fitosanitario y gradiente de infección de la enfermedad escoba de bruja *Crinipellis perniciosa* (Stahel) Singer. Ingeniero Agrónomo Thesis, University of Caldas.

Lagos, G. & Guarnizo, N. (1991). Control fitosanitario y gradiente de infección de la enfermedad escoba de bruja *Crinipellis perniciosa* (Stahel) Singer. Ingeniero Agrónomo Thesis, University of the Llanos.

Ecuador

Arias M., R. (1990). La escoba de bruja del cacao: influencia de fuentes de inoculo vecinas. Ingeniero Agrónomo Thesis, University of Cuenca.

Lara P., E. (1991). Estudio sobre la gradiente de infección de escoba de bruja y la relación escoba de bruja y monilia en cacao (*Theobroma cacao* L.). Ingeniero Agrónomo Thesis, University of Guayaquil.

Macias C., G.E. (1988). La humedad sobre los tejidos y su relación can la epifitología de la escoba de bruja en cacao tradicional. Ingeniero Agrónomo Thesis, University of Guayaquil.

Moreira S., C.J. (1989). Aspectos epidemiológicos de la escoba de bruja *Crinipellis perniciosa* en la zona de Quevedo. Ingeniero Agrónomo Thesis, Technical University of Manabi.

---

[2]One student from Viçosa has not yet completed his M.S. thesis.

Trinidad

Mohan S. (1991). Epidemiology of witches' broom disease of cocoa (*Theobroma cacao* L.) in Trinidad. M. Phil. Thesis, University of the West Indies.

Appendix IV

IWBP EXPERIMENTAL PROTOCOLS

PROTOCOL 1 : COMPARATIVE EPIDEMIOLOGY EXPERIMENT

1. Passport Data

Sites should be chosen, representative of farmers' cocoa plantations with witches' broom, and the following recorded:

1.a Plantation

A representative plantation will be selected, according to the interest of the participants, and the following recorded:

1. Area
2. Soil
3. Shade
4. Tree age
5. Spacing
6. Plot location on map
7. Distance to other cocoa (plantations and/or wild cocoa)
8. History
9. Ground cover and litter
10. Cultural practices

1.b Plot

Ten representative sample trees will be chosen to form a plot within the plantation. These trees are to be selected in an area containing a minimum of 50 trees. They are assigned and marked with a number.

The area of the plantation containing the plot is to be shown on a plan indicating shade and other trees, extent of shade canopy and position of site meteorological station.

1.c Trees

The following are recorded for each of the 10 marked trees:

1. Maximum height (m)
2. Trunk girth at 0.5 m
3. Jorquette height/number
4. Canopy width/depth

5. Genetic material
6. Any peculiarities of the tree

2. Marking of trees and collection of data sets

2.a Vegetative

Five units each of 10 healthy growing points are selected on each numbered tree representative of the canopy as a whole.

The following are recorded on these units every 2 weeks:

1. Number of flush points
2. Total number of flush leaves per unit
3. Number of:
   - (i) Terminal brooms
   - (ii) Axillary brooms
   - (iii) Stem swellings/cankers
   - (iv) Swellings pulvinus/petiole

Each event, (i) – (iv), is marked when first observed and labelled with date (week number) and the initial symptom.

2.b Cushions

Four 1 m lengths on branches throughout the canopy and one 1 m length in the middle of the trunk (50 cm if the trunk is short) are selected, marked and numbered on each tree. The following are recorded within these lengths every 2 weeks:

1. Number of active cushions (active = when flower buds appear);
   When cushion first shows activity it is marked with a pin.
2. Number of cushions with open flowers.
3. Number of cushions with presence/absence of:
   - (i) dead flowers
   - (ii) chirimoyas
   - (iii) broom

   Each cushion is marked with a different coloured pin to denote (i), (ii), or (iii) when each event is first observed.

2.c Fruit

The following are recorded every 2 weeks on the five 1 m lengths designated for the cushion data sets and five additional 1 m lengths in the branches of the tree:

1. Number of newly-set fruit > 1 cm long. Each is marked with pin to indicate date observed.
2. Number of fruit wilted.
3. Number of healthy ripe fruit harvested.
4. Number of pods with necrosis. These are removed, counted (carrot shaped pods/other pods) and each is split to confirm; in later observations presence of *Crinipellis, Moniliophthora, Phytophthora* should be noted.
5. Number of fruits newly recorded as chirimoyas. These are not removed.
6. Number of pods with green islands.

For categories 2, 3 and 4 the date of fruit set is recorded. All other diseased and ripe pods on the trees in the plot are removed and counted every 2 weeks. The height of each pod to the nearest 50 cm is recorded on the 10 sample trees.

2.d Basidiocarp Production

Sporulation is recorded as open, turgid basidiocarps on one day in each calendar week, if possible on a day after rain. Data are recorded early in the morning. Basidiocarps are not removed. Where appropriate observations are made throughout the year.

Basidiocarps are counted within the following units:

1. Suspended Vegetative Brooms
   - (i) On the samples of brooms hung in the tree in year 1. When these brooms fall they are not replaced.
   - (ii) On each numbered tree a new representative sample of 10 numbered dead brooms hung at the top of the canopy and another sample of 10 numbered, dead brooms hung in the middle of the canopy. These brooms are obtained within the plantation. If marked brooms fall they are re-hung if possible or, if not, are replaced.
2. Attached Vegetative Brooms
   Attached brooms formed in year 1 up to a maximum of 10 brooms per tree within the marked units.
3. Brooms on Ground
   - (i) On brooms placed beneath each numbered tree in year 1.
   - (ii) A new sample of 10 numbered dead fallen brooms placed beneath each tree in year 2. (At some sites only, an additional sample of 10 numbered pruned brooms).
4. Cushions
   - (i) On a maximum of 20 infected cushions per tree identified in year 1. The remaining cushions infected in year 1 are retained. Fallen infected material is not replaced.
   - (ii) On each marked tree, where they are present 20 infected cushions distributed on the tree. In situations where they are not present, then on trees outside the plot, 20 per tree where available. If this option is not possible, on cushion brooms obtained from outside the site and attached to marked trees to a maximum of 20 per tree.
5. Pods
   On each numbered tree, a representative sample of 10 infected pods hung in the canopy, 5 at the top and 5 in the middle, alongside the suspended brooms, and 10 samples of infected pods from the plantation, placed beneath the numbered trees (10 per tree). Fresh samples of pods are added in year 2.
6. Other Sources
   Checked, and recorded if sporulation occurs. Fresh samples of pods are added in year 2. If an object is lost (L) or replaced (R) it should be noted in the appropriate position on the data sheets in the week in which it occurs.

2.e Once the units designated for collection of data are marked on each numbered tree, vegetative brooms, cushion brooms and infected pods are removed from the remainder of the tree and counted.

Diseased tissue is removed on other trees within the plot, and this material is left on the ground in the plot.

2.f Meteorological

1. Meteorological data are available from established stations near the designated plots.
2. The following equipment is placed outside the canopy of trees, in an open space, at each site:
   (i) Casella Thermohygrograph
   (ii) Weighing bucket rainguage
   (iii) Belfort dew recorder
   (iv) Maximum and minimum thermometer
   (v) Sling psychrometer

   The thermohygrograph, maximum/minimum thermometer and sling psychrometer are placed in a weather shelter at standard meteorological height. The dew recorder is placed near the weather shelter at the same height as the thermohygrograph.
3. Parameters required from these sources are:

(i)	Temperature	maximum
		minimum
		hours > 30°C
		hours < 20°C
(ii)	Relative humidity	maximum
		minimum
		hours > 95 %
		hours < 70 %
(iii)	Rain – each event	start
		finish
		duration
		amount
(iv)	Wetness – each event	start of any deflection
		finish of any deflection
		maximum % deflection recorded
		duration of 100 % deflection
		cause of event (rain and/or dew).

   Day of reference is day event begins.

2.g Sampling weeks begin on Mondays and end on Sundays.
For newly established sites the observations made every two weeks should start in an odd–numbered week.

PROTOCOL 2 : 'COMPARATIVE PHYTOSANITATION EXPERIMENT'

1. Site and Experimental Design

1.a A cocoa planting of not less than 0.5 ha, at least 250 m and preferable 500 m from other cocoa is divided into two areas, A and B, consisting of 10 rows and a minimum of 35 rows respectively. Twenty trees in each row, ten either side of a central axis areas A and B constitute the experimental site.

1.b The following are recorded:

1. Geographic location
2. Latitude/longitude
3. Elevation above sea level (m)
4. Topography
5. Soil
6. Shade
7. Planting material, age and height for the experimental site
8. Spacing
9. Map of planting, showing experimental site and distance to other cocoa

1.c The experiment is maintained for a minimum of 27 months, including the 'calibration' year.

2. Agronomic Practices

Any necessary structural pruning of the whole cocoa planting should be completed in month 1. Existing cultural practices, including chupon removal, will be maintained throughout the <u>experimental site</u> after month 1, but no more structural pruning will be done.

In the area outside the experimental site, all normal cultural practices will be continued.

3. 'Calibration'

After approximately 12 months of site 'calibration' (pod harvests, etc.) all infected material is removed from Area B. Area A remains a source area without removal.

4. Treatment

4.a In Area B, all infected material is removed after 12 months and 24 months, with secondary removals after 15 months and 27 months. The timing of the secondary removals can be adjusted to accord with local practice. Infected material is removed from the planting and destroyed.

4.b A treatment, e.g. chemical spray, can be applied to one–half of the trees (10 trees) in each row of the experimental site within Area B provided this maintains the area as one in which all infected material is removed.

5. Collection of Data

5.a <u>Fruit</u>

Numbers of harvested fruit, collected every 2 weeks, separated into various classes:

ripe/healthy; presence of *Crinipellis, Moniliophthora, Phytophthora*, others. Records are kept for each individual tree within the experimental site.

5.b <u>Brooms</u>

When infected material is removed, the numbers of vegetative infections and cushion infections are recorded on six trees (randomly selected at the start of the experiment) in each row of the experimental site within Area B (three trees from the treated and non-treated halves of the row).

At these times, vegetative infections and cushion infections are removed from 10 marked trees (randomly chosen, one from each row) of the experimental site within Area A, and counted.

Immediately following the removal in month 15, 20 trees randomly chosen from those on which removal counts were made within Area B and two trees of the 10 marked trees within Area A are inspected branch by branch and remaining infections counted. Infections are classified within the categories used in Protocol 1 for flush and flower data sets.

5.c <u>Basidiocarps</u>

Numbers of open, turgid basidiocarps are recorded on one day in each calendar month (if possible on a day after rain) on two collections of 100 brooms each suspended in the middle of the canopy as five samples of 20 brooms on five, randomly selected trees within the experimental site of Area A. One collection of brooms is randomly selected, the other collection is of productive brooms.

REFERENCES

Allen, J.B. (1987). London Cocoa Trade Amazon Project. Final Report Phase 2. *Cocoa Growers' Bulletin* **39**, 1–94.

Allen, J.B. & Cabrera, J.J. (1988). Early evaluation of the INIAP–London Cocoa Trade germplasm collection in Ecuador. *Proceedings of the 10th International Cocoa Research Conference, Santo Domingo, Dominican Republic, May 1987*, pp. 679–684.

Allen, J.B. & Lass, R.A. (1983). London Cocoa Trade Amazon Project. Final Report Phase 1. *Cocoa Growers' Bulletin* **34**, 1–71.

Almeida, H.A., Machado, R.C., Villa Nova, N.A. & da Silva, W.S. (1987). Influência de elementos meteorológicos no lançamento do cacaueiro. *Revista Theobroma* **17**, 163–174.

Almeida, L.C. de (1988). Sobrevivência do *Crinipellis perniciosa* durante a fermantação do cacau. *Proceedings of the 10th International Cocoa Research Conference, Santo Domingo, Dominican Republic, May 1987*, pp. 359–362.

Almeida, L.C. de & Andebrhan, T. (1988). Recuperação de plantações de cacau com alta incidência de vassoura-de-bruxa na Amazonia Brasileira. *Proceedings of the 10th International Cocoa Research Conference, Santo Domingo, Dominican Republic, May 1987*, pp. 337–339.

Alvares-Afonso, F.M. (1984). *A cacauicultura da Amazônia, antecedentes, estruturas programáticas, evolução e resultados alcançados.* CEPLAC, Brasilia, Brazil, 85pp.

Alvim, P. de T. (1977). Cacao. In *Ecophysiology of Tropical Crops*, ed. P. de T. Alvim & T.T. Koslowski. Academic Press, New York, pp. 279–313.

Alvim, P. de T., Machado, A.D. & Vello, F. (1974). Physiological responses of cacao to environmental factors. *Revista Theobroma* **4**, 3–25.

Andebrhan, T. (1981). Relação entre idade do fruto do cacaueiro e suscetibilidade a *C. perniciosa*. *Informe Tecnico 1981. CEPLAC/DEPEA*. Belém, Pará, Brazil, pp. 69–70.

Andebrhan, T. (1985*a*). Studies on the epidemiology and control of witches' broom disease of cacao in the Brazilian Amazon. Proceedings of the 9th International Cocoa Research Conference, Lome, Togo, February 1984, pp. 395–402.

Andebrhan, T. (1985*b*). Relação entre sombreamento e nivel de infecção por *C. perniciosa* em cacaueiros. *Informe Tecnico 1985. CEPLAC/DEPEA.* Belém, Pará, Brazil, pp. 81–84.

Andebrhan, T. (1985*c*). Recuperação de cacauais com alto nivel de infecção causada por *C. perniciosa. Informe Technico 1985. CEPLAC/DEPEA.* Belém, Pará, Brazil, pp. 92–93.

Andebrhan, T. (1988*a*). Cultural characteristics and biochemical reactions of isolates of *Crinipellis perniciosa*, causative agent of witches' broom disease of cocoa. *Proceedings of the 10th International Cocoa Research Conference, Santo Domingo, Dominican Republic, May 1987*, pp. 363–366.

Andebrhan, T. (1988*b*). Rain water as a factor in the dissemination of basidiospores of *Crinipellis perniciosa* (Stahel) Singer within cacao trees. *Proceedings of the 10th International Cocoa Research Conference, Santo Domingo, Dominican Republic, May 1987*, pp. 367–369.

Andebrhan, T. & Almeida L.C. de (1985). Recuperação de cacauais com alto nivel de infecção causada por *C. perniciosa. Informe Tecnico 1985. CEPLAC/DEPEA*. Belém, Pará, Brazil, pp. 92–93.

Andebrhan, T. & Almeida, L.C. de (1987). Relação entre concentração de basidiosporos de *C. perniciosa* e infecção em *T. cacau. Informe Tecnico 1987. CEPLAC/DEPEA.* Belém, Pará, Brazil, pp. 66–68.

Anon (1992). *Cocoa Market Report, September 1992* **No. 394**. E. D. & F. Man, London, U.K.

Aragundi, J.A. (1974). Evaluación de rendimiento e incidencia de enfermedades del cacao en varias zonas ecológicas del Litoral Ecuatoriano. Ingeniero Agrónomo Thesis, University of Guayaquil, Ecuador.

Aragundi, J.A. (1982). Culture and histology of *Crinipellis perniciosa* (Stahel) Singer. M.S. Thesis, University of Florida.

Aragundi, J.A., Frias, G., Solorzano, G., Schmidt, R.A., & Purdy, L.H. (1988). Estudios sobre gradientes de infección y dispersión de la escoba de bruja del cacao en el Ecuador. *Proceedings of the 10th International Cocoa Research Conference, Santo Domingo, Dominican Republic, May 1987*, pp. 375–379.

Aranzazu, F. (1982*a*). Algunos aspectos de la biologia de *Crinipellis perniciosa* (Stahel) Singer en la region de Urabá (Colombia). *Proceedings of the 8th International Cocoa Research Conference, Cartagena, Colombia, October 1981*, pp. 323–328.

Aranzazu, F. (1982*b*). Analisis de la evolucion y variacion de la incidencia de escoba de bruja (*Crinipellis perniciosa*) en la region de Urabá (Colombia). *Proceedings of the 8th International Cocoa Research Conference, Cartagena, Colombia, October 1981*, pp. 387–393.

Aranzazu, H., F. (1986). Determinación de la frecuencia de remoción de escoba en cultivos de la zona marginal baja cafetera. *Proceedings 7th ASCOLFI Congress of Phytopathology, Paipa, Colombia*. (Abstract).

Arias, R. (1990). La escoba de bruja del cacao: influencia de fuentes de inóculo vecinas. Ingeniero Agrónomo Thesis, University of Cuenca, Ecuador.

Baker, R.E.D. (1953). Witches' broom disease of cocoa (a review). *A Report on Cocoa Research 1945–1951*, pp 116–118.

Baker, R.E.D. & Crowdy, S.H. (1943). Studies in the witches' broom disease of cacao caused by *Marasmius perniciosus* Stahel. Part I. Introduction, symptoms and etiology. *Department of Mycology and Bacteriology Memoir No. 7*. Imperial College of Tropical Agriculture, St. Augustine, Trinidad.

Baker, R.E.D. & Crowdy, S.H. (1944). Studies in the witches' broom disease of cacao caused by *Marasmius perniciosus* Stahel. Part II. Field studies and control methods. *Department of Mycology and Bacteriology Memoir No. 8*. Imperial College of Tropical Agriculture, St. Augustine, Trinidad.

Baker, R.E.D. & Dale, W.T. (1944). Witches' broom disease investigations. X. Loss of pods in I.C.S. clones at River Estate during 1943. *Tropical Agriculture Trinidad* **21**, 196–198.

Baker, R.E.D. & Holliday, P. (1957). Witches' broom disease of cacao (*Marasmius perniciosus* Stahel). *Phytopathological Paper* **No. 2**, 42 pp. Commonwealth Mycological Institute, Kew, Surrey.

Baker, R.E.D. & McKee, R.K. (1943). Witches' broom disease investigations. VI. The infection of flower cushions and pods of cacao by *Marasmius perniciosus* Stahel. *Tropical Agriculture, Trinidad* **20**, 188–194.

Baker, R.E.D., Crowdy, S.H. & Thorold, C.A. (1941). Witches' broom disease investigations. I. Seasonal variations in the intensity of infections and their effect on control measures. *Tropical Agriculture Trinidad* **18**, 107–116.

Bartley, B.G. (1977). The status of genetic resistance in cacao to *Crinipellis perniciosa* (Stahel) Singer. *Proceedings of the 6th International Cocoa Research Conference, Caracas, Venezuela, November 1977*, pp. 57–69.

Bastos, C.N. (1982). Influência da temperatura na liberação e germinação de basidiosporos de *Crinipellis perniciosa* (Stahel) Singer. *Informe Técnico 1980. CEPEC/CEPLAC*. Itabuna, Bahia, Brazil, pp. 307–308.

Bastos, C.N. (1984). Efeito de filtrado de cultura de *Cladobotryum amazonense* sobre *Crinipellis perniciosa* (Stahel) Singer e outros patogenos. *Revista Theobroma* **14**, 263–269.

Bastos, C.N. & Andebrhan, T. (1986). Urucu (*Bixa orellana*): nova especie hospedeira da vassoura-de-bruxa (*Crinipellis perniciosa*) do cacaueiro. *Fitopatologia Brasileira* **11**, 963–965.

Bastos, C.N. & Evans, H.C. (1979). Resultados preliminares sobre o estabelecimento de um sistema de controle da vassoura de bruxa na Amazonia. *Comunicado Técnico* **No. 12**. *CEPLAC/DEPEA*. Belém, Brazil.

Bastos, C.N. & Evans, H.C. (1985). A new pathotype of *Crinipellis perniciosa* (witches' broom disease) on solanaceous hosts. *Plant Pathology* **34**, 306–312.

Bastos, C.N., Evans, H.C. & Samson, R.A. (1981). A new hyperparasitic fungus, *Cladobotryum amazonense*, with potential for control of fungal pathogens of cocoa. *Transactions of the British Mycological Society* **77**, 273–278.

Bastos, C.N., Neill, S.J. & Horgan, R. (1986). Effect of *Cladobotryum* antibiotic on the major pathogens of cacao in Brazil. *Revista Theobroma* **16**, 213–221.

Bastos, C.N., Andebrhan, T. & Almeida, L.C. de (1988). Comparação morfologica de isolados de *Crinipellis perniciosa*. *Fitopatologia Brasileira* **13**, 202–206.

Bravo, E.V. & Hedger, J.N. (1988). Microflora associated with "witches broom" in cocoa and its potential role in the biological control of the pathogen *Crinipellis perniciosa*. *Proceedings of the 10th International Cocoa Research Conference, Santo Domingo, Dominican Republic, May 1987,*. pp. 345–348.

Calle, H. (1978). Relation of internal infection by *Crinipellis perniciosa* to witches' broom formation in cacao. M.S. Thesis, University of Florida.

Calle, H., Cook, A.A. & Fernando, S.Y. (1982). Histology of witches' broom caused in cacao by *Crinipellis perniciosa*. *Phytopathology* **72**, 1479–1481.

Capriles de R., L. (1973). Recientes estudios llevados a cabo en Venezuela en torno a la enfermedad escobas de bruja del cacao *(Marasmius perniciosus)*. *Proceedings of the International Conference on Cocoa Diseases, Guayaquil, Ecuador, 1973*. (Abstract)

Capriles de R., L. (1977). *Enfermedades del cacao en Venezuela*. Fondo Nacional del Cacao, Venezuela, 82 pp.

Capriles de R., L. (1983). Jornadas Técnicas Estación Experimental Caucagua. Caucagua, Venezuela.

Capriles de R., L. (1988). El control quimico dentro del manejo integral de la escoba de bruja (*Crinipellis perniciosa*) del cacao. *I Jornadas Técnicas de la Asociación de Procesadores de Cacao*. Colonia Tovar, Venezuela.

Capriles de R., L. & Freites, A. (1977). Efectividad de fungicidas para el control de la enfermedad escoba de bruja *(Crinipellis perniciosa)* en plantas de cacao. *IX Jornadas Agronómicas*. Maracay, Venezuela.

Capriles de R., L. & Reyes, E.H. (1962). Situación fitosanitaria del área cacaotera de la Victoria – Estado Apure. *I Jornadas Nacionales de Cacao*.

Capriles de R., L. & Rojas, E. (1977). Comportamiento de cultivares de cacao al ataque de la enfermedad escoba de bruja *(Crinipellis perniciosa)*. *IX Jornadas Agronómicas*. Maracay, Venezuela.

Carzola, I.M., Aidar, T. & Milde, L.C.E. (1989). Perfis do lançamento, floração, da bilração e de estágios do fruto do cacaueiro no Estado da Bahia no período de 1977/78. *Boletim Técnico, CEPLAC/CEPEC*. Itabuna, Brazil, 58 pp.

Cifuentes, C., Mayorga, M., Prieto, E. Rondon, G. & Tovar R., G. (1982*a*). Estudio de la infeccion y esporulacion del *Crinipellis perniciosa* (Stahel) Singer, agente causal de la escoba de bruja del cacao en Guamal Colombia. *Proceedings of the 8th International Cocoa Research Conference, Cartagena, Colombia, October 1981,* pp. 484–492.

Cifuentes, C., Mayorga, M., Prieto, E. Rondon, G. & Tovar R., G. (1982*b*). Estudio quantitavo de la produccion de escobas en plantas de cacao, afectadas por *Crinipellis perniciosa* (Stahel) Singer y su significado en el manejo de la enfermedad. *Proceedings of the 8th International Cocoa Research Conference, Cartagena, Colombia, October 1981*, pp. 407–413.

Cox, A.E. & Large, E.C. (1960). Potato blight epidemics throughout the world. *Agricultural Handbook No. 174*. Agricultural Research Service, United States Department of Agriculture, Washington D.C., 230 pp.

Cronshaw, D.K. (1979). Fungicide application together with cultural practices to control cocoa diseases in Ecuador. *Tropical Agriculture, Trinidad* **56**, 165–170.

Cronshaw, D.K. & Evans, H.C. (1978). Witches' broom disease of cocoa (*Crinipellis perniciosa*) in Ecuador. II. Methods of infection. *Annals of Applied Biology* **89**, 193–200.

Cruikshank, A.M. (1970). Cocoa in Grenada. *Cocoa Growers' Bulletin* **15**, 4–11.

Cubillos A., G. & Grisales, A. (1985). Rehabilitación de poblaciones de cacao deterioradas por escoba de bruja, mediante el método de "renovación de copa". *El Cacaotero Colombiano* **31**, 36–41.

Danquah, O–A. (1986). Growth of *Crinipellis perniciosa* in cocoa resistant and susceptible to witches' broom. Ph.D. Thesis, University of London.

Delgado, J.C. (1974). Studies of *Marasmius perniciosus*. Ph.D Thesis, University of Florida.

Delgado, J.C. & Cook, A.A. (1976). Nuclear condition of the basidia, basidiospore and mycelium of *Marasmius perniciosus*. *Canadian Journal of Botany* **54**, 66–72.

Desrosiers, R. & Bolaños, C.W. (1955). Inhibition of sporophore development of *Marasmius perniciosus* (Stahel) on cacao. *Turrialba* **5**, 28–32.

Edwards, D.F. (1978). Studies on the manipulation of the timing of crop maturity of cocoa in Ecuador in relation to losses from pod diseases. *Journal of Horticultural Science* **53**, 243–254.

Evans, H.C. (1978). Witches' broom disease of cocoa (*Crinipellis perniciosa*) in Ecuador. I. The fungus. *Annals of Applied Biology* **89**, 185–192.

Evans, H.C. (1980). Pleomorphism in *Crinipellis perniciosa*, causal agent of witches' broom disease of cocoa. *Transactions of the British Mycological Society* **74**, 515–523.

Evans, H.C. (1981). Witches' broom disease - a case study. *Cocoa Growers' Bulletin* **32**, 5–19.

Evans, H.C. & Solorzano, G. (1982). Witches' broom disease: wrong experiment, right results. *Proceedings of the 8th International Cocoa Research Conference, Cartagena, Colombia, October, 1982*, pp. 415–418.

Evans, H.C., Edwards, D.F. & Rodríguez M. (1977). Research on cocoa diseases in Ecuador: past and present. *Pest Abstracts and News Summaries* **23**, 68–80.

Fonseca, S.E.A. (1988). Assessing resistance to *Crinipellis perniciosa* in cocoa clones and callus cultures. Ph.D Thesis, University of London.

Fonseca, S.E.A. & Wheeler, B.E.J. (1990). Assessing resistance to *Crinipellis perniciosa* using cocoa callus. *Plant Pathology* **39**, 463–471.

Fonseca, S.E.A., Almeida, L.C. de & Andebrhan, T. (1985). Patogenicidade de isolados e avaliação de resistência de clones de cacau a *Crinipellis perniciosa*. *Proceedings of the 9th International Cocoa Research Conference, Lome, Togo, February 1984*, pp. 233–236.

Frampton, V.L., Linn, M.B. & Hansing, E.D. (1942). The spread of virus diseases of the yellows type under field conditions. *Phytopathology* **32**, 799–808.

Freeman, W.G. (1928). Witch-broom in Trinidad. *Tropical Agriculture Trinidad* **5**, 287–288.

Frias, G.A. (1987). An inoculation method to evaluate resistance to witches' broom disease of cocoa. Ph.D Thesis, University of Florida.

Giraldo, L. & Valencia, J. (1985). Evaluación de dos fungicidas en el control de la escoba de bruja de cacao. Ingeniero Agrónomo Thesis, University of Caldas, Colombia.

Gregory, P.H. (1948). The multiple-infection transformation. *Annals of Applied Biology* **35**, 412–417.

Gregory, P.H. (1968). Interpreting plant disease dispersal gradients. *Annual Review of Phytopathology* **6**, 189–212.

Gregory, P.H. (1973). *Microbiology of the atmosphere*. 2nd Edition. Leonard Hill, London.

Griep, D.N. & Lima, E.L. (1988). Aumento da eficiência técnica e econômica pela eliminação ou substituição de cacaueiros de baixa produtividade. *Informe de Pesquisa. CEPLAC/DEPEA/COPES*. Belém, Para, Brazil, pp. 120–122.

Grisales, A. & Cubillos, G. (1985). Rehabilitacion de poblaciones de cacao deterioradas por escoba de bruja mediante el metodo de "renovacion de copa". *El Cacaotero Colombiano* **31**, 36–41.

Hardy, F. (1954). *Cacao manual* (English ed.). Interamerican Institute of Agricultural Sciences, Turrialba, Costa Rica, 395 pp.

Hedger, J.N. (1985). Tropical agarics: resource relations and fruiting periodicity. In *Developmental Biology of Higher Fungi*, ed. D. Moore, L.A. Casselton, D.A. Wood & J.C. Frankland. Cambridge University Press, Cambridge, pp. 41–86.

Hedger, J.N. (1988) Decomposition and basidiocarp production by witches' brooms of cocoa under field conditions in Ecuador. *Proceedings of the 10th International Cocoa Research Conference, Santo Domingo, Dominican Republic, May 1987*, pp. 319–323.

Hedger, J.N., Pickering, V. & Aragundi, J. (1987). Variability of populations of the witches' broom disease of cocoa (*Crinipellis perniciosa*). *Transactions of the British Mycological Society* **88**, 533–546.

Hernández, A. (1958). *Bases para la realización de una campana de renovación cacaotera*. Ministerio de Agricultura y Cria, Venezuela.

Holliday, P. (1954). Control of witches' broom disease of cacao in Trinidad. *Tropical Agriculture, Trinidad* **31**, 312–317.

Holliday, P. (1970). *Crinipellis perniciosa. C.M.I. Descriptions of Pathogenic Fungi and Bacteria* **No. 223**. Commonwealth Mycological Institute, Kew.

Humphries, E.C. (1942). Dormancy of cocoa buds. *11th Annual Report Cocoa Research, I.C.T.A. Trinidad*, pp. 28–38.

INIAP (1972). Distancias de siembra en Cacao Híbrido. *Informe Anual, 1971*. Instituto Nacional de Investigaciones Agropecuarias, Estación Experimental Tropical Pichilingue.

INIAP (1974*a*). Podas en Cacao Híbrido. *Informe Anual, 1973*. Instituto Nacional de Investigaciones Agropecuarias, Estación Experimental Tropical Pichilingue.

INIAP (1974*b*). Fertilización intensiva. *Informe Anual, 1973*. Instituto Nacional de Investigaciones Agropecuarias, Estación Experimental Tropical Pichilingue.

INIAP (1975). Uso de aceite agricola para inhibir producción de basidiocarpos. *Informe Anual, 1974*. Instituto Nacional de Investigaciones Agropecuarias, Estación Experimental Tropical Pichilingue.

Jeger, M.J. (1983). Analysing epidemics in time and space. *Plant Pathology* **32**, 5–11.

Khonje, P.T. (1984). Effects of environmental factors on an isolate of *Crinipellis perniciosa* (Stahel) Singer, the cause of witches' broom disease of cocoa (*Theobroma cacao* L.). M.Sc. Thesis, University of Reading.

Kiyosawa, S. & Shiyomi, M.A. (1972). A theoretical evaluation of the effect of mixing resistant variety with susceptible variety for controlling plant disease. *Annals of the Phytopathological Society of Japan* **38**, 41–51.

Kranz, J. (ed.) (1974). *Epidemics of Plant Disease: Mathematical Analysis and Modeling*. Springer–Verlag, Berlin, Heidelberg, New York.

Kranz, J. (1978). Comparative anatomy of epidemics. In *Plant Disease: Advanced Treatise. Vol. II. How Disease Develops in Populations*, ed. J.G. Horsfall & E.B. Cowling. Academic Press, New York, 436 pp.

Krupasagar, V. & Sequeira, L. (1969). Auxin destruction by *Marasmius perniciosus*. *American Journal of Botany* **56**, 390–397.

Lachenbruch, P.A. & Goldstein, M. (1979). Discriminant analysis. *Biometrics* **35**, 69–85.

Laker, H. & Rudgard, S.A. (1989). A review of the research on chemical control of witches' broom disease of cocoa. *Cocoa Growers' Bulletin* **42**, 5–16.

Laker, H.A., Sreenivasan, T.N. & Rajkumar, D. (1988*a*). Present status of witches' broom disease of cocoa in Trinidad. *Tropical Pest Management* **34**, 318–323.

Laker, H.A., Sreenivasan, T.N. & Rajkumar, D. (1988*b*). The resistance of some cocoa clones to *Crinipellis perniciosa*. *Proceedings of the 10th International Cocoa Research Conference, Santo Domingo, Dominican Republic, May 1987*, pp. 637–641.

Lara, E. (1991). Estudio sobre la gradiente de infección de escoba de bruja y la relación escoba de bruja y monilia en cacao, *Theobroma cacao* L.. Ingeniero Agrónomo Thesis, University of Guayaquil, Ecuador.

Lass, R.A. & Rudgard, S.A. (ed.) (1987). *Report of the 4th Workshop of International Witches' Broom Project, 11–17 September 1987, Belém, Brazil*. International Office of Cocoa Chocolate and Sugar Confectionery, Brussels, Belgium.

Lindeberg, G. & Molin, K. (1949). Notes on the physiology of the cocoa parasite, *Marasmius perniciosus*. *Physiologia Plantarum* **2**, 138–144.

Macillas, G.E. (1988). La humedad sobre los tejidos y su relación con la epifitología de la escoba de bruja en cacao tradicional. Ingeniero Agrónomo Thesis, University of Guayaquil, Ecuador.

Maddison, A.C. (1985). *Informe Técnico, 1981–85, Estación Experimental Tropical Pichilingue*. Instituto Nacional de Investigaciones Agropecuarias, Quito, Ecuador, pp. 6–7.

Maddison, A.C. (1991). *Informe Técnico 1986–1990. Estacion Experimental Tropical Pichilingue*. Instituto Nacional de Investigaciones Agropecuarias, Quito, Ecuador, pp. 9–14.

Maddison, A.C. & Mogrovejo, E. (1984). Witches' broom and the establishment of cacao. *Proceedings of the 9th International Cocoa Research Conference, Lomé, 1984*, pp. 389–394.

Materán, M. (1981). Estudio de diferentes aislados del hongo (*Crinipellis perniciosa*). *X Jornadas Agronómicas*. San Cristóbal, Venezuela.

Mayorga, M. (1984). *Informe anual de actividades, Programa de Fitopatología*. Instituto Colombiano de Agropecuaria, Bogotá, Colombia.

Mayorga, M. (1989). Host–parasite relationships between isolates of *Crinipellis perniciosa* and cocoa hybrids from Colombia. M.Phil. Thesis, University of London.

McGeary, F.M. & Wheeler, B.E.J. (1988). Growth rates of, and mycelial interactions between, isolates of *Crinipellis perniciosa* from cocoa. *Plant Pathology* **37**, 489–498.

McKee, R.K. (1942). Some observations on I.C.S. clonal material at River Estate. *11th Annual Report Cocoa Research, I.C.T.A. Trinidad*.

Minogue, K.P. (1986). Disease gradients and the spread of disease. In *Plant Disease Epidemiology: Population Dynamics and Management Vol. 1*, ed. K.J. Leonard & W.E. Fry. Macmillan, New York, pp. 285–310.

Moreira, C.J. (1989). Aspectos epidemiologicos de la escoba de bruja, *Crinipellis perniciosa* en la zona de Quevedo. Ingeniero Agrónomo Thesis, University of Manabí, Ecuador.

Muse, R.B., Isaac, S., Collin, H.A. & Hardwick, K. (1988). Phytotoxin production by *Crinipellis perniciosa*. *Proceedings of the 10th International Cocoa Research Conference, Santo Domingo, Dominican Republic, Santo Domingo, Dominican Republic, May 1987*, pp. 355–358.

Orchard, J.E. & Hardwick, K. (1988). Photosynthesis, carbohydrate translocation and metabolism of host and fungal tissues in cacao seedlings infected with *Crinipellis perniciosa*. *Proceedings of the 10th International Cocoa Research Conference, Santo Domingo, Dominican Republic, May 1987*, pp. 325–

330.

Orchard, J.E., Hardwick, K., Collin, H.A., Isaac, S. & Dart, J. (1989). The role of plant growth regulators in the response of cocoa to infection by *Crinipellis perniciosa*. *Report for the Biscuit, Cake, Chocolate & Confectionery Alliance, August 1989.*

Palti, J. & Kranz, J. (ed.) (1980). *Comparative Epidemiology: A Tool for Better Disease Management*. Center Agricultural Publications Document, Wageningen, Netherlands, 122 pp.

Parra, D. (1989). *Informe Anual*. Fondo Nacional de Investigaciones Agropecuarias, Venezuela.

Pegler, D.N. (1978). *Crinipellis perniciosa* (Agaricales). *Kew Bulletin* **32,** 731–736.

Pegus, J.E. (1972). Aspects of the host-parasite relationships in the *Theobroma cacao L./Marasmius perniciosus* Stahel disease complex. M.Sc. Thesis, University of the West Indies, Trinidad.

Pereira, J.L. (1985). Possible strategy in chemical control of witches' broom disease of cocoa. *Report of the 2nd Workshop of the International Witches' Broom Project, Wageningen, October 1985.* ed. R.A. Lass & S.A. Rudgard. International Office of Cocoa Chocolate and Sugar Confectionery, Brussels, Belgium, pp. 10–20.

Pereira, J.L., Ram, A., Figueiredo, J.M. & Almeida, L.C. de (1989). Primeira ocorrência de vassoura-de-bruxa na principal região produtora de cacau do Brasil. *Revista Agrotrópica* **1,** 79–81.

Pérez, Z.A., Reyes, E.H. & Capriles de R., L. (1972). Programa de mejoramiento genético del cacaotera en Venezuela. *Proceedings of 4th International Cocoa Research Conference, St Augustine, Trinidad.*

Posnette, A.F. & Palma, M. (1944). Observations on cacao at Paria Peninsula, Venezuela. *Tropical Agriculture* **21,** 131.

Pound, F.J. (1938). Cacao and witchbroom disease of South America. *Report on a visit to Ecuador, the Amazon Valley and Colombia.* Department of Agriculture, Trinidad and Tobago, 58 pp.

Purdy, L.H., Trese, A.T. & Aragundi, J.A. (1983). Proof of pathogenicity of *Crinipellis perniciosa* to *Theobroma cacao* by using basidiospores produced in *in vitro* cultures. *Revista Theobroma* **13,** 157–163.

Pyke, E.E. (1933). The physiology of cacao. 1. General observations on growth flowering and fruiting. *2nd Annual Report Cocoa Research, I.C.T.A. Trinidad,* pp. 37–40.

Reyes, E.H. (1963). Selecciones de cacao Playa Alta nueva fuente de germoplasma para resistencia a escoba de bruja (*Marasmius perniciosus*) del cacao. *Boletin Cacao.*

Reyes, E.H. (1973). *Informe anual de gestión.* Fondo Nacional de Investigaciones Agropecuarias, Venezuela.

Robinson, R.A. (1976). *Plant Pathosystems.* Springer-Verlag, Berlin.

Robinson, R.A. (1987). *Host Management in Crop Pathosystems.* Collier-Macmillan, London.

Rocha, H.M. (1983). The ecology of *Crinipellis perniciosa* (Stahel) Singer in witches' brooms on cocoa (*Theobroma cacao* L.). Ph.D Thesis, University of London.

Rocha, H.M. & Wheeler, B.E.J. (1982). The water balance as an important factor in basidiocarp production by *Crinipellis perniciosa*, the causal fungus of cocoa witches' broom. *Proceedings of the 8th International Cocoa Research Conference, Cartagena, Colombia, October 1981,* pp. 381–386.

Rocha, H.M. & Wheeler, B.E.J. (1985). Factors influencing the production of basidiocarps and the deposition and germination of basidiospores of *Crinipellis perniciosa*, the causal fungus of witches' broom on cocoa (*Theobroma cacao*). *Plant Pathology* **34,** 319–328.

Rodriguez, G.R. (1988). Resultados de la transferencia de technologia en el combate de escoba de bruja (*Crinipellis perniciosa*) del cacao en el Alto Beni, Bolivia. *Proceedings of the 10th International Cocoa Research Conference, Santo Domingo, Dominican Republic, May 1987,* pp. 751–754.

Rondon, G. (1986). Control cultural de escoba de bruja y moniliasis en cacao. *Procedimientos del 7to ASCOLFI Congreso de Fitopatologia, Paipa, Colombia.* (Abstract).

Rorer, J. B. (1926). Ecuador cacao. *Tropical Agriculture, Trinidad* **3,** 46–47 and 68–69.

Rudgard, S.A. (1986*a*). Witches' broom disease of cocoa in Rondonia, Brazil: pod losses. *Tropical Pest Management* **32,** 24–26.

Rudgard, S.A. (1986*b*). Witches' broom disease on cocoa in Rondonia, Brazil: basidiocarp production on detached brooms in the field. *Plant Pathology* **35,** 434–442.

Rudgard, S.A. (1987). Interpreting the epidemiology of cocoa witches' broom for better disease management in Rondonia, Brazil. *Cocoa Growers' Bulletin* **38,** 28–38.

Rudgard, S.A. (1989). International Witches Broom Project. Detailed description of symptoms of witches' broom disease of cocoa caused by *Crinipellis perniciosa*. *Cocoa Growers' Bulletin* **41,** 32 pp.

Rudgard, S.A. & Andebrhan, T. (1988). Predicting the cost-benefits of sanitation pruning for the management of witches' broom disease. *Proceedings of the 10th International Cocoa Research Conference, Santo Domingo, Dominican Republic, May 1987*, pp. 341–344.

Rudgard, S.A. & Butler, D.R. (1987). Witches' broom disease on cocoa in Rondonia, Brazil: pod infection in relation to pod susceptibility, wetness, inoculum and phytosanitation. *Plant Pathology* **36**, 515–522.

Rudgard, S.A. & Lass, R.A. (1985). Some economic information on cocoa production in South American countries affected by witches' broom disease. In *Report of the 2nd Workshop of the International Witches' Broom Project, Wageningen, October 1985*, ed. R.A. Lass & S.A. Rudgard. International Office of Cocoa Chocolate and Sugar Confectionery, Brussels, Belgium, pp. 23–29.

Sá, D.F., de Almeida, M.A., da Silva, L.F. & Leão, A.C. (1982). Fatores edafoclimáticos seletivos ao zoneamento da cacauicultura no Sudeste da Bahia. *Revista Theobroma* **12**, 169–187.

Scerne, R.M.C. & Santos, M.M. (1987). Fenologia do cacaueiro na Amazônia. *Informe Técnico 1987. CEPLAC/DEPEA*. Belém, Pará, Brazil, pp. 16–19.

Singer, R. (1942). A monographic study of the genera *Crinipellis* and *Chaetocalathus*. *Lilloa* **8**, 503.

Singh, J. (1937). Escoba de bruja del cacao (*Marasmius perniciosus*). *Boletin de la Asociación Venezolana de Productores de Cacao* **2**, 51–52.

Solorzano, G.R. (1977). Factores ambientales involucrados en la produccion de basidiocarpos y en la diseminacion de basidiosporas de *Marasmius perniciosus* Stahel. Tesis, Ingenerio Agronomo, Technical University of Manabi.

Soria, J. (1970). The latest cocoa expeditions to the Amazon basin. *Cacao, Turrialba* **15**, 5–15.

Sreenivasan, T.N. & Dabydeen, S. (1989). Modes of penetration of young cocoa leaves by *Crinipellis perniciosa*. *Plant Disease* **73**, 478–481.

Sreenivasan, T.N., Petitt, T.R. & Rudgard, S.A. (1990). An alternative method of applying copper to control cocoa pod diseases. *Proceedings of Brighton Crop Protection Conference Pests and Diseases, 1990*, pp. 583–588.

Stahel, G. (1915). *Marasmius perniciosus* Nov. Spec. *Bulletin Departement Van den Landbouw in Suriname* **No. 33**, 25 pp. [Translated from the German by A.M.W. Ter Laag]

Stahel, G. (1919). Bijdrage tot de Kennis der Krullotenziekte. *Bulletin Departement van den Landbouw in Suriname* **No. 39**. [Contribution to the knowledge of witch broom disease. Translation by B.G. Montserin (1932) in *Tropical Agriculture, Trinidad* **9**, 167–176]

Stell, F. (1928). Witch-broom disease of cacao and its control. *Bulletin Department Agriculture, Trinidad & Tobago*, **No. 21**.

Stell, F. (1937). *Administrative Report Department Agriculture, Trinidad & Tobago, 1936*, pp 52–56.

Suárez, C., C. (1977). Growth of *Crinipellis perniciosa* (Stahel) Singer *in vivo* and *in vitro*. Ph.D Thesis, University of London.

Suárez, C., C. (1987). Enfermedades del cacao y su control. In *Manual del cultivo de cacao*, ed. C. Suárez C., Estación Experimental Tropical Pichilingue, Quevedo, Ecuador. pp. 70–80.

Thorold, C.A. (1943). Witches' broom investigations. V. *Tropical Agriculture, Trinidad* **20**, 176–181.

Thorold, C.A. (1975). *Diseases of Cocoa*. Clarendon Press, Oxford.

Thresh, J.M. (1976). Gradients of plant virus diseases. *Annals of Applied Biology* **82**, 381–406.

Tovar R., G. (1991). La escoba de bruja del cacao (*Crinipellis perniciosa* (Stahel) Singer) en el piedemonte llanero de Colombia: epidemiologia, resistencia y manejo. *Agronomia Colombiana* **8**, 1–239.

Toxopeus, H. (ed.) (1981). *Archives of Cocoa Research. Volume I. Cocoa populations and witches' broom disease in South America*. American Cocoa Research Institute, USA / International Office of Cocoa Chocolate and Sugar Confectionery, Belgium.

Vanderplank, J.E. (1963). *Plant Diseases: Epidemics and Control*. Academic Press, New York.

Wheeler, B.E.J. (1985*a*). Witches' broom disease (*Crinipellis perniciosa*) of cocoa and its quarantine significance. *Proceedings of the Regional Conference on Plant Quarantine Support for Agricultural Development, December 1985*. ASEAN Plant Quarantine & Training Institute, Selangor, Malaysia, pp. 13–18.

Wheeler, B.E.J. (1985*b*). The growth of *Crinipellis perniciosa* in living and dead cocoa tissue. In: *Developmental Biology of Higher Fungi*, ed. D. Moore, L.A. Casselton, D.A. Wood & J.C. Frankland. Cambridge University Press, Cambridge, pp. 103–116.

Wheeler, B.E.J. (1987). Plant pathology in a developing world. British Society for Plant Pathology Presidential Address 1986. *Plant Pathology* **36**, 430–437.

Wheeler, B.E.J. & Mepsted, R. (1982). Pathogenic races of *Crinipellis perniciosa? Proceedings of the 8th International Cocoa Research Conference, Cartagena, Colombia, 1981*, pp. 365–370.

Wheeler, B.E.J. & Mepsted, R. (1988). Pathogenic variability amongst isolates of *Crinipellis perniciosa* from cocoa (*Theobroma cacao*). *Plant Pathology* **37**, 475–488.

Wood, G.A.R. (1985). Production. In *Cocoa*, ed. G.A.R. Wood & R.A. Lass. Longmans, London, pp. 543–586.

Wood, G.A.R. & Lass, R.A. (1985). *Cocoa*. Fourth Edition. Longmans, London.

INDEX

MIX
Papier aus verantwortungsvollen Quellen
Paper from responsible sources
FSC® C105338

If you have any concerns about our products,
you can contact us on
ProductSafety@springernature.com

In case Publisher is established outside the EU,
the EU authorized representative is:
Springer Nature Customer Service Center GmbH
Europaplatz 3, 69115 Heidelberg, Germany

Printed by Libri Plureos GmbH
in Hamburg, Germany